THE LIARS OF NATURE AND THE NATURE OF LIARS

LIXING SUN

The Liars *of* Nature *and the* Nature *of* Liars

CHEATING AND DECEPTION IN THE LIVING WORLD

PRINCETON UNIVERSITY PRESS

PRINCETON AND OXFORD

Published by Princeton University Press
41 William Street, Princeton, New Jersey 08540
99 Banbury Road, Oxford OX2 6JX

press.princeton.edu

All Rights Reserved

Library of Congress Control Number 2022948596
ISBN 978-0-691-19860-6
ISBN (e-book) 978-0-691-24573-7

British Library Cataloging-in-Publication Data is available

Editorial: Alison Kalett and Hallie Schaeffer
Production Editorial: Jill Harris
Text Design: Heather Hansen
Jacket Design: Heather Hansen
Production: Danielle Amatucci
Publicity: Sara Henning-Stout and Kate Farquhar-Thomson
Copyeditor: Susan Matheson

Jacket image: Leaf mimic katydid (Family *Pseudophyllidae*), Morley Read/ Alamy Stock Photo.

This book has been composed in Arno Pro

Printed on acid-free paper. ∞

Printed in the United States of America

10 9 8 7 6 5 4 3 2

For Shine and Orien
and their world in the future

CONTENTS

THE LIARS OF NATURE AND THE NATURE OF LIARS

Liar, Liar, Everywhere

She is pregnant. Raising a child takes a lot of time and energy, yet she is short of both. Homeless, she has no choice but to find somebody else to take care of her baby—for free. It's not easy, but she knows how to pull it off. She scouts around and spots a cozy house in a quiet neighborhood. The young wife of the family looks caring and has just given birth to a new baby, so is a perfect choice as a surrogate. She hides herself and waits in the vicinity, keeping watch on the house. Opportunity presents itself when the new mother takes a short trip to get some food. She sneaks in and switches the baby with her own. Then she heartlessly throws the victim's infant in a dump.

What you have just read is a cold-blooded murder case, one that takes place in nature when a female cuckoo bird sneaks her egg into a warbler's nest. The cuckoo is cheating, though the scenario doesn't quite fit *Oxford English Dictionary*'s definition of the verb "cheat": to "act dishonestly or unfairly in order to gain an advantage." Cheating in humans usually involves an element of intention. In the larger biological world, however, establishing intent is neither easy nor necessary. For biologists, as long as organisms act to favor themselves at the expense of others—especially in situations when cooperation is expected—they are cheating.[1]

This book is about the behavior, evolution, and natural history of cheating. Although, in common usage, the word "cheating" is often interchangeable with "lying" and "deceiving," the three words differ in connotation, nonetheless. Furthermore, lying and deceiving involve two very different biological processes, as we will unveil in the next two chapters. In light of this new insight, the word "cheating" refers to both lying and deceiving in the book.[2]

🜚

Cheaters are everywhere in the biological world, according to our broadened definition of cheating. Monkeys sneak around for sex; possums, well, play "possum," as they are famous for when pursued by a predator; birds scare rivals away from contested food by crying wolf—emitting alarm calls that are normally used to warn others about an approaching predator; amphibians and reptiles are master impostors, altering their body color to blend into their backgrounds; stickleback fish protect their eggs and babies by misdirecting their cannibal peers away from their nests; defenseless caterpillars ward off predators by masquerading as dangerous animals such as snakes with big false eyes (see color plate 1); squids escape from predators by ejecting ink to create a "smoke screen" in the water. Examples of lying and deceiving behavior in animals can go on and on.

What may surprise you is that cheating doesn't require a brain, or even a neuron, as many plants are cheaters as well. For example, most orchids mimic the aromas of their pollinators' food. Around 400 orchid species, however, evolved a more audacious tactic: they fool male pollinators by mimicking the smell and appearance of female insects to take advantage of eager males who seek opportunities to mate (see color plate 2). Even more amazing, these plants can keep male pollinators aroused by preventing them from ejaculating. Thus, the unsatisfied male pollinators will keep going in search of another female—including a female-apparent flower—to mate with. Since these males are highly promiscuous, they are extremely effective in spreading orchid pollen.[3]

Fungi cheat too. For example, truffles—mushroom-like species that form fruiting bodies underground—emit a steroid called androstenol that mimics the pheromone of wild boars. Androstenol is produced in the testes of adult boars and has a musty odor to the human nose. When female pigs sense the truffle aroma, they will dig exuberantly for the source. What they don't know is that they are being suckered by something bearing no resemblance to the swine beau they are hoping for. The only outcome of their passionate fervor is spreading spores for the truffles.[4] Mission accomplished for the fungi that deceive.[5]

Complex organisms such as plants and fungi cheat; so does single-celled life. A good example is the slime mold (or social amoeba) known by its scientific name, *Dictyostelium discoideum* (or "Dicty" for short). When starved, the slime mold amoeba cells gather together to form a mobile, slug-like structure. The "slug" moves as a unit until it finds a suitable spot and then grows into a fruiting body made of a spore-producing head mounted on a thin stalk. The entire thing is shaped like a lollipop or a maraca (a rattle-like percussion instrument popular in Latin America) (fig. 1.1). The cells in the head, which consists of 80% of all cells, will seed the next generation when food becomes plentiful again. The other 20% of the cells consigned to the stalk, however, rot away after completing their mission—to raise the head so that the spores can scatter far and wide, like dandelions spreading their fluffy seeds in the wind.

If you were a slime cell, where would you prefer to end up—the head or the stalk of the fruiting body? The head, of course! Because only in the head do you have the opportunity to pass your genes to the next generation. If you were a cell in the stalk, your genes would be destined for an evolutionary dead end. Who, in the biological world, wants to be relegated to an inferior status, without a chance to reproduce?

Fortunately, this isn't a major issue when amoeba cells have the same genetic makeup, like identical twins. When cells share an identical set of genes, it matters little as to which cells seed the next generation. However, when a fruiting body is made of a chimera of two or more types of cells, where many of the genes are different, conflict ensues. They all compete to be part of the fertile head rather than play a supporting role in the sterile stalk. As one might expect, different cells play dirty in order to make it into the prized head by any means necessary, including cheating.[6] Some types of cells, enabled by certain genetic mutations, defraud others by sending more than their fair share of "representatives" to the head,[7] a maneuver similar to political gerrymandering. Moreover, once they have made it to the head, they produce noxious chemicals to

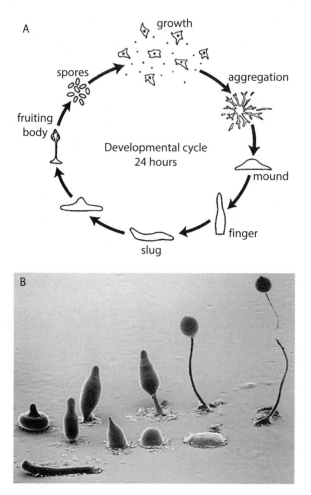

FIGURE 1.1. The developmental cycle (A) and sequence of spore formation (B) in the social slime mold *Dictyostelium discoideum* (Myre 2012).

prevent latecomers from getting into the lifeboat for the next generation. Recent studies have revealed that more than 100 mutant genes are implicated in this amoeba cheating scam.[8]

Next, let's visit the bacterial world to see whether they also cheat. Bacteria are tiny. Individuals alone can't do much. Just as building the Great Wall involved hundreds of thousands of humans, achieving collective bacterial tasks (such as emitting light—bioluminescence—and trapping vital elements from the environment) requires millions of bacteria working together toward a common goal. That's why bacteria often congregate to form a thin, slimy layer called biofilm in soil or water.

Among their communal projects is gathering iron, an element critical for bacteria to survive. The problem they face is that iron is usually found in low concentrations in their surroundings. Because individual bacteria can't do much on their own, collecting iron is necessarily a collaborative work for the community. To coordinate their efforts, bacterial members "talk" with one another by releasing chemicals that signal particular genes to "turn on" in sync to make a family of complex compounds called siderophores. Siderophores are related to hemoglobin in our blood cells in that they can bind with iron. In this way, they serve as bacteria's fishing net to scoop up iron floating in the environment.

But there is a catch. Siderophores are costly to produce for individual bacteria in terms of material and energy. Yet they are a public good, a "commons" shared by all members in the community. As we all know, once you have a public good, there are often cheaters who come along for a free ride. Who hasn't been in a situation where some team members take credit for the group's work yet contribute less than other members?

Bacteria also suffer from this social dilemma. There is no shortage of free riders that contribute less than their fair share to the production of siderophores yet still devour the catch—iron—along with all the others who've done the work.[9] Obviously, cheaters can sabotage the collective effort. If they are too numerous, productivity of siderophores will drop, and the amount of iron collected will decline, which in turn will put the livelihood of the entire community at risk. Threatened by this potentially fatal consequence, honest producers have evolved an arsenal of antifraud strategies. Some bacteria, for example, can band together with their genetic look-alikes to prevent free riders from infiltrating their community. They may even use toxins to kill the cheaters.[10]

Even viruses cheat. Viruses are not considered fully alive, because they lack the necessary biological tools to survive and reproduce on their own. They have to steal their hosts' resources and genetic machinery to complete their life cycles. This means that to engage in cheating does not require a complete form of life.

Blatant cases of viral cheating have been well-documented. One example occurs when different viral species or different variants of the same species infect a single host cell. Their biological resources—such as genes and proteins—can become mixed. This provides opportunities for some viruses to use trickeries to steal the resources produced by others that serve as unwilling helpers. In this way, cheating viruses don't need to possess all the genes essential for making copies of themselves or for assembling their protein coatings, known as capsids, that package their genetic material within.[11]

What we have seen so far are cases of cheating between different individuals, simple or complex, unicellular or multicellular. Cheating can also take place within the same individual. Cancer cells, for example, are cheating cells that shirk the duty of cooperation with other cells in the body. They instead gobble up all the resources, proliferate, and refuse to commit suicide when commanded to do so. Thus, fighting cancer is essentially fighting cheating cells, a point made clear in Athena Aktipis's 2020 book, *The Cheating Cell*.

Even inside a typical cell, cheating is part of life. For instance, the B chromosome ekes out a living by cheating. In sharp contrast to the normal "A" chromosomes that are familiar to us, B chromosomes are smaller and can be common, and they can be found in varying numbers within a cell (fig. 1.2). What makes them stand out is their ability to tag along without doing any work. In other words, they are passed on, generation after generation, by hitchhiking without contributing to cell functionality, akin to party crashers living on free food while enjoying themselves at their host's expense.

Even genes cheat. Your body is a vessel for a vast amount of genetic garbage known as junk DNA. Just like B chromosomes, junk DNA serves no purpose for its host organism but gets a free ride from generation to generation.[12] The amount of junk DNA is truly awe-inspiring. It accounts for up to 98% of our genome, encompassing many varieties of useless genetic material, such as repeated elements, pseudogenes, and transposable elements (the last of which are more vividly called jumping genes).

Jumping genes are fragments of DNA that can insert themselves almost anywhere in the genome through a copy-and-paste process, much as we do in a word processing program. They are so prolific that they make up 45% of the entire human genome.[13] A well-known jumping gene is the *Alu* element. With a length of about 300 base pairs, the *Alu* element has multiplied more than a million copies of itself in the evolutionary lineage leading to humans over the last 53 million years. Today, it makes up 10.7% of the entire human genome.[14] Because of the hy-peractive nature of jumping genes and their ability to promote them-selves, the genome of salamanders can be 40 times larger than that of humans.[15] Since all animals (in fact, all eukaryotic organisms) have roughly the same number of working genes, what salamanders can boast about is an extremely large genetic junkyard.

FIGURE 1.2. B chromosomes (marked with "B") in the roe deer (modified from Graphodatsky et al. 2011).

As befits their name, jumping genes jump, and they jump randomly in the genome. Because the vast majority of our DNA is junk, whenever a jumping gene replicates itself and places the copy in a new location in our genome, it usually has no noticeable effect, like adding a new bag of trash in a huge landfill. But once in a while such an insertion may hit an area in the middle of a functional gene. If this happens, it can result in a serious genetic defect and lead to health problems such as cancer or hemophilia.[16]

Intrigued by jumping genes? There are also bizarre cases of cheating genes called selfish genetic elements or, more colorfully, outlaw genes. Among the most famous are genes known as segregation distorters or meiotic drivers in insects. In the common lab fruit fly, *Drosophila mela-nogaster*, these genes can boost their own representation in the genome by killing sperm cells that carry alternative alleles. In doing so, these outlaw genes get more than their fair share of what would normally be an even split.[17] If these outlaw genes are located on the X or Y chromosome, it may result in a lopsided sex ratio (more males or females), rather than a 50:50 divide.[18]

The last case of cheating genes I want to describe are known as converting elements. These genes code for enzymes called homing endonucleases, which can cut open a DNA chain at specific locations. They then add a replica of themselves into the breach.[19] It's like a rogue physician who uses his own sperm to inseminate eggs from women who come to him seeking artificial fertilization.

Converting elements commit a genetic fraud by violating the rules followed by other genes. The genes that play by the rules may be hurt directly by becoming disabled, or indirectly by being outcompeted in an unfair race. Like jumping genes, converting elements can be transmitted horizontally, copying and inserting themselves into the genome of their peers in addition to their own descendants. (Unexpectedly, this "rogue" quality of self-promotion has given a new halo to homing endonucleases today: it's the foundation for the technology of gene editing called CRISPR, pioneered by Jennifer Doudna and Emmanuelle Charpentier, who together won the 2020 Nobel Prize in Chemistry.)

Selfish genetic elements such as B chromosomes, jumping genes, segregation distorters, and converting elements share a common feature: they all promote their own interests at the expense of other genes. Since the transmission patterns of these genetic elements violate classical Mendelian laws, you may feel that you learned the wrong things in high school biology class. But no worry. Biological systems are complex and are rarely governed by universal laws as in physics. Because of this, biology is known for being a science of exceptions.

※

The above section, although citing only a few examples, demonstrates that cheating is found in all domains of life, at every level of the biological hierarchy, from the most complex organisms to the least sophisticated, even incomplete, forms of life. It is found among animals, plants, fungi, bacteria, viruses, chromosomes, genes, and snippets of

DNA. It occurs within the same individual, between individuals of the same species, and between species that are vastly different in form and function.

Regardless of their prevalence in nature, however, the words cheating, lying, and deception all come with negative connotations due to our moral preference and the premium we place on honesty. Although we value truth and loathe lies, real life often runs counter to what we ideally want. Contrary to the long-held dictum, honesty is not always the best policy in our daily lives.

Consider this case. An innocent man has been falsely charged, convicted, and condemned to die. Desperate to save him, his loyal friends propose a way out: escape by bribing the jailer. Even when faced with this choice, however, he declines on the ground that to do so would be cheating the legal system. What do you think about the concept of honesty as applied by this man? If you were in his position, what would you do?

If you think the man's choice is foolish, congratulations! You've just saved the life of Socrates, the Greek philosopher who chose death over breaching the trust between a citizen and the state. How likely is it that we would find a heroic martyr, willing to die for the sake of trust and honesty, in the natural world? Extremely unlikely—in fact, no known examples exist. On the contrary, we find that cheating is ubiquitous in nature at all levels.

Why is cheating so common in the biological world? The answer: evolution is not a Socratic philosopher. It is, instead, an unmoral, heartless process that proceeds pragmatically without any concern over ethical preferences, honor codes, or value systems. It certainly makes no distinction between prosocial cooperation and antisocial manipulation, because all that matters is what works to enhance survival and reproduction. Any trait—be it morphological, physiological, behavioral, or genetic—can prevail as long as it can boost its owner's Darwinian fitness, defined and measured as the number of offspring born and raised to adulthood. Furthermore, while freeing cheating from our moral consideration, evolution punishes those who forgo it as a strategic option when using it can increase their fitness. As a result, even

though it might seem brazen and despicable to our human social sensi-
bilities, cheating thrives in the biological world.

So, cheating flourishes in nature as a direct result of natural selection.
Less well-known, however, is that cheating also serves as a potent
selective force that drives evolution on its own. The reason is simple in
concept: cheating favors the cheater and hurts the cheated. As such, it
spurs the emergence of counter-cheating tactics, which in turn beget
counter-counter-cheating strategies, ad infinitum. And during this on-
going evolutionary arms race, to quote Darwin, "endless forms most
beautiful and most wonderful have been, and are being, evolved."

To illustrate this point, take cheating in rhizobia, soil bacteria that
live in the roots of plants—specifically legumes. These bacteria fix nitrogen for plants, whereas the plants provide housing facilities and food
in the form of carbon. So, the relationship is supposed to be happily
mutualistic—or so we traditionally thought. But a close examination
has revealed that, rather than a love affair, the relationship between rhizobia and their plant hosts is far more complicated. Some rhizobia actually produce very little nitrogen. That is, they cheat in order to get free
housing and carbon from the plants.[20] For this reason, not all plants
welcome rhizobia. Some are known to fight back by cutting off the nu-
trient supply if cheating rhizobia are too numerous. Only those living
in poor soil, desperately in need of nitrogen, would grudgingly put up
with an unfair relationship with rhizobia.[21] Apparently, beggars can't be
choosers. This demonstrates how cheating can unleash a cascade of new
moves and countermoves as the bacteria and their hosts try to get the
upper hand in their relationship.

Intrigued by the complex strategies emerging from the evolutionary
game played by rhizobia and plants? This is just a simple case to illustrate how cheating can trigger an evolutionary arms race and become a
powerful catalyst for the creation of diversity, complexity, and even
beauty, as we will see in the following chapters.

Unfortunately, the role of cheating in evolution remains underap-
preciated today for two key reasons. One is historical. Darwin himself
did not address cheating as a major force in evolution by natural selection. *On the Origin of Species* never mentions the word "cheat" but

uses the word "deceive" seven times. Only three are related to animal cheating—all are forms of mimicry, protective disguises employed by tasty bugs to fool their predators. Clearly, how cheating relates to evolution and biodiversity wasn't on his mind—at least not high in priority among his many ideas.

Darwin's omission implies the second reason for us to overlook the importance of cheating. It's easy to see natural selection in terms of unrelenting, cutthroat competition for resources between rivals, or in terms of surviving the onslaughts of predators, parasites, and pathogens. Because of this, evolution has been popularly stereotyped as "survival of the fittest" and "nature red in tooth and claw." Such a one-dimensional impression tends to divert our attention from the soft power of cooperative behaviors that are fully as effective for enhancing fitness in numerous situations and contexts, a point made clear by many scientists during recent decades.

In some animals, social intelligence is significantly more important than physical strength. In a bonobo group, for instance, success in terms of fitness is predicated on the strength of an individual's social network. A brawny brute who relies on sheer individual muscle power is destined to be a loser when faced with the united efforts of cooperating members of the group. Without some needed social intelligence, he could also become an object of manipulation, exploited by others. This is why cheating, a catalyst for social intelligence, matters so much in evolution.

𝕩

With the emergence of modern human intelligence, the arms race between cheating and counter-cheating strategies was not only vastly expanded and intensified but also began to take place at a whole new level—the arena of cultural evolution. And just as it results in the emergence of novel biological traits, cheating is a potent catalytic force that spurs many cultural innovations, which then lead to cultural diversity and complexity. Without cheating, there would be no literature, art, science, technology, business, or religion—and the list goes on until it encompasses all aspects of our lives, society, and culture. This may seem deeply counterintuitive for now, but the reasons will become evident later when we zero in on how modern technologies and cultural institutions evolve and transform in sync with cheating.

Despite my emphasis on the catalytic power of cheating, I have no desire or intent to create a revisionist account of the virtue of lies. On the contrary, many forms of cheating, whether or not they are considered criminal offenses, can cause substantial harm to innocent people. That's why no serious moral philosophy or religion would endorse or advocate cheating. As social scientists have amply shown us, the basic cultural glue that binds us together as a human society is trust.[22] Though it may not seem so from what I've portrayed in this chapter, this book will reinforce this point from a biological perspective. It's unthinkable that a society could sustain itself for long without honesty and truth as its moral foundations. That's why we humans have fought so hard to suppress cheating and cheaters for millennia.

Yet, despite our best efforts, cheating has been a persistent and perennial problem in all known human societies across history. No society, in fact, has ever succeeded in completely wiping it out. Moreover, as if the problem were not bad enough, cheating has become perceptibly worse in the Information Age. Not only do all traditional trickeries continue to exist, but cheating has extended itself into the digital realm, where it has found a fertile environment to thrive. A vast number of new scams, from phishing to sextortion, are emerging and evolving with increasing sophistication and reach. At the societal level, the ubiquity of fake news and conspiracy theories poses a major threat to democracy. By preventing citizens from acquiring accurate and reliable information, it undermines our ability to agree on the basic nature of truth and fact.[23] What should we do about cheating, given that we are unable to root it out?

The seemingly quixotic and fatalistic campaign against cheating doesn't mean that it's not worth trying, nor that we're doomed to lose. Rather, it gives us an opportunity to rethink our approach in this digital age and find new ways to deal with a problem that has been with us since time immemorial. In this respect, evolutionary science can offer a trove of wisdom for us to tap into.

This book will provide a tour into the world of cheaters to see how organisms use a broad spectrum of methods to deceive, hustle, and swindle others for their own gain. More importantly, we'll search for the modus operandi behind the vast diversity of tricks, scams, and frauds. We will then use our newly acquired evolutionary understanding of how cheaters operate to design novel strategies to combat cheating in our society.

Specifically, we will, in the next two chapters, journey into how animals cheat by using two rules that carry through the book. We will then find reasons for how and why honesty can survive and flourish amid the onslaught of lies and deceptions in chapter 4. With this information in mind, we will turn to chapter 5 to see how cheating can spur the emergence of novel features including behavior, intelligence, and art, all through evolutionary arms races. This will be followed by two chapters about human cheating and self-deception, respectively, showing that the rules used in cheating apply in both the biological realm and human cultures. Finally, we will venture into philosophical terra incognita, attempting to settle the age-old controversy: Is there any cheating that is morally acceptable?

When you close the book, I hope you'll be convinced by the book's main premise and overarching goal: cheating is a powerful catalyst that contributes to the creation of diversity, complexity, and beauty in the biological and cultural world. By understanding how it works, cheating can be practically contained, even though it may seem like an inevitable and invincible part of life.

Now, buckle up for a thrilling journey into the world of cheating.

CHAPTER 2

Hackers and Suckers in Communication

Imagine you're a hungry crow, working hard to find a meal on a cold winter day. Tired from several failed hunting trips, you fly to a tree branch to rest. Just after settling on a perch, you see dozens of your peers fighting raucously over the remains of a dead skunk on the snowy road below. You realize that you're late for the party and that the notorious corvid bully is also on the scene. What can you do to get a bite? One good solution would be to cry wolf and scare all of them away. When your rivals panic and rush for safety, you can swoop down and snatch a chunk of the skunk.

The above scenario is a description of crow behavior in the real world. More importantly, the scenario embodies a general rule about how animals cheat: they falsify honest information in communication for their own interest, which is the biological essence of lying. Let's add some fun and call this the *First Law of Cheating*. (Yes, there is a Second Law of Cheating, to be introduced in the next chapter.) Before we delve into how animals apply this law to cheat in a practical sense, we need to address a more basic issue: Why do animals cheat at all? The answer lies, paradoxically, in their need to cooperate.

Cooperation is a good thing. It can allow two animals to reap greater benefits by working together rather than going it alone. The benefits of cooperation apply to all living entities including plants, fungi, and bacteria, but our focus here is on animals. So, whenever animals intermingle, they have a strong incentive to work together. To do so, however, they need to talk—to communicate. Therefore, communication is a necessary precondition for cooperation to take place.[1] Once communication evolved, however, theatrical drama also tagged along as an inevitable by-product.

As it takes two to tango, a basic communication loop is made of signals transmitted between a sender and a receiver. As simple as the system is, dramatic things can still happen during the process. Imagine yourself onstage as the sender. You face two options: lie or tell the truth. Whichever option you choose, you hope to gain more from your interaction with the receiver than you would by pursuing the same action alone. Otherwise, there's no reason to communicate. Producing and sending signals takes time and energy; it may also draw unwanted attention from predators or parasites. When shutting up and staying put is better for you, why waste your resources and risk your life to send a message to somebody else?

For the sender, therefore, to lie or not to lie is a matter of an evolutionary imperative, not a Shakespearean moral choice. To be adaptive, the sender should be a Machiavellian manipulator, unencumbered by any moral consideration. This sender-as-manipulator idea was championed by Richard Dawkins and John Krebs, two Oxford biologists who, in the 1970s, cut out a logical path through a jungle of muddled thoughts regarding the nature of animal communication.[2]

The theatricals don't stop after a signal is sent, of course. The receiver on the other end also faces a dilemma upon picking up a message from the sender: to respond or not to respond. What should it do? To answer this question, imagine yourself as the CEO of a widget company. One day, you receive a proposal from a potential partner, claiming that a new joint venture can raise your productivity from 2 units per hour to 3 units per hour. If you let it pass, you'll lose an opportunity to raise your profits by 50%. Yet when you perform a risk analysis by crunching Big Data, it tells you that there is also a 34.21% chance that you're being scammed. If this is the case, your productivity will drop to 1 unit per hour, and your profit be cut in half—a giant loss. Should you accept the proposal or not?

With the information available, you calculate that if you accept the bid, your productivity is expected to be 2.32 units per hour after factoring in the risk.[3] This is 16% better than your current level. Although it's not the 50% increase your prospective partner claims, the increase in productivity still seems too good to pass up.

In much the same way, many animals are bombarded by tempting proposals from their peers and are forced to make a choice. Although they don't produce widgets like pins, planes, or cell phone plans, they have an equally important stake on the line: their Darwinian fitness. (Recall from chapter 1 that Darwinian fitness is the number of offspring they can produce and raise to adulthood.) How do they decide? Although they can't do a cost-benefit analysis by crunching numbers, they can do it by following their instincts—software implemented in their sensory and cognitive system by a powerful agent named evolution.

Just as in your deciding whether to accept the unsolicited proposal for your company, animals can use their cognitive capacities to calculate potential gains and losses before they make a choice. If on average the benefit outweighs the cost, they should respond to the signal positively. Otherwise, they'd be better off simply ignoring it. If they do so, the communication loop will collapse.[4]

The consequences of communication collapse are illustrated by a tiny critter: the Pacific field cricket. In this species, males come in two types—either honest chirpers or quiet sneakers—in their strategies for seeking mating opportunities. Chirping males sing courtship songs to females, broadcasting their eagerness to mate. They do so by rubbing their serrated wings, like scraping two files against each other. Sneakers, however, have smooth wings and are unable to produce sound. So, they sit near singing males and ambush females who are looking for a date. Because females prefer chirpers, they keep the population of sneaking males down to a small minority, which is only a by-product of their occasional indiscretion.

In Hawaii, where the cricket was introduced, singing males met their nemesis—a parasitic fly that can eavesdrop on their songs to advance the fly's own interests. The female fly can locate male crickets by their chirps and lays eggs on them. After hatching, fly maggots burrow into the cricket's body and hold an extended feast on its guts—until the cricket dies.

On the island of Kauai, biologist Marlene Zuk and her fellow researchers were stunned by the power of evolution in action: in just a decade beginning in 1991, the number of singing males took a nosedive, from abundant to extremely rare. When they dug around, however, the researchers still found quite a few crickets. But these crickets were different. Nearly all the males were mute with smooth wings—they were originally the sneakers.[5] The honest chirpers had been decimated by the parasitic fly. As cheaters came to dominate the population, the advantage of the acoustic channel for cricket dating disappeared—the result of being hacked by a parasitic fly.

This case illustrates a key point: animal communication, as testified by its very existence, must be rooted in some fundamental level of trust among those in the loop. That is, for a communication loop to be sustained, senders must be reasonably honest, and receivers must be on average better off if they respond to signals rather than if they ignore them. This is why, as biologists believe, communication in animals is mostly cooperative.[6] Cheating, by and large, is a nuisance in the overwhelming triumph of cooperation facilitated by communication.

During communication between two individuals, how does lying succeed? In most situations, senders cheat by altering the meaning of their signals. That is, they falsify the message. When the intended receivers interpret a false message as being true to its literal meaning, they end up being defrauded. That's exactly how the fabled boy-who-cried-wolf was initially able to pull off his lie: people responded as if real wolves were coming.

To further illustrate the point, let's put into perspective the lying crow scenario from the beginning of the chapter. When a crow spots a stalking fox, it caws an alarm sound to its peers, alerting them *a fox is coming*! The message is truthful. But if a crow makes the same call when there is no danger, the message in the signal is false. If others react as if the false message were truthful by fleeing the scene, they've been duped. Cheaters are exploiting the element of assumed honesty in communication to gain a fitness advantage. In other words, cheaters falsify truthful information to benefit themselves at the expense of their peers. This is how the First Law of Cheating works.

In the rest of this chapter, we'll take a tour into a diverse arsenal of tactics animals use to manipulate information. Before we move on, we need to make a distinction between two kinds of communication: *within species* and *between species*. The remainder of this chapter will focus on within-species communication. We'll reserve between-species communication for the next chapter.

♊

When it comes to cheating, there are two key questions to ask: What do cheaters want? And how do they get it? The first is relatively easy to answer. Animals, like us, cheat in order to promote their own Darwinian fitness. That's how all organisms prosper in evolution by natural selection. Unsurprisingly, fitness enhancers such as food, social status, and opportunities to mate are on the most-wanted list of items for which animals will cheat. The second question—how they get what they want—is more difficult to answer because, for every goal, there are multiple pathways, epitomized by the old, albeit gruesome, proverb, "There's more than one way to skin a cat." Let's take a look at some of the details to see what patterns will emerge.

Crows are far from alone in their penchant to fake alarm calls to cheat for food. Tits, flycatchers, and many other birds use the same technique. Since evolution often proceeds as an arms race, stealing food by cheating can lead to countermeasures by victims who can also lie to protect their food. This locks victims and victimizers into an information war, where both sides try to outmaneuver the other party.

Crows, ravens, and jays, for example, are always aware of their peers' presence. They often keep their distance or deliberately misdirect rivals from food sources known only to themselves. Misdirection is even more important for lower-ranking individuals. Often, it's the only option they have to protect their food. Subordinate ravens, for instance, keep the locations of food sources to themselves. When necessary, they employ tactics to trick dominant individuals, causing them to search in areas away from the subordinates' own caches of food. Only when the subordinates feel certain of a head start will they visit their secret storage to feast. But like monkeys, ravens are too smart to be easily fooled. The misled birds can quickly figure out they've been duped and begin searching for the hidden food instead of blindly following their lying peers.[7]

Squirrels are not to be outdone by birds in devising cunning schemes to protect their food. Not only do they do most of the same things birds do, but they also take extra steps to conceal information about their food. When they're busy storing food, for instance, they keep a distance from their peers. If this is impossible, they'll turn their back toward others so as not to be caught in the act. In case others come nosing around, squirrels may fabricate fake caches as decoys to protect their food.[8] Out of curiosity, I often watch gray squirrels hoarding acorns. However, when I go to check what they've been hiding, I often find empty holes—I've been fooled.

Similar behaviors are found in macaques, squirrel monkeys, vervet monkeys, and capuchin monkeys. These primates live in small hierarchical societies where higher-ranking members are known to rudely snatch favorite foods from those lower on the rung. To avoid this, lower-ranking members regularly withhold information about food.[9] The same behavior has also been observed in chimpanzee groups—lower-ranking members avoid revealing the locations of prized food (such as a hidden banana) when a higher-ranking member is around. They refuse to go anywhere near the banana unless they think it's safe.[10]

In birds and mammals, false alarms are not only used for food; they can also be used for sex. For instance, barn swallows will give off alarm calls when they spot a predator. During the breeding season, they nest together so that there are more eyes watching for predators, which benefits all the members in the colony. The downside is that a communal life also facilitates illicit affairs when neighbors live side-by-side—all eager to mate. In order to safeguard their paternity—to protect against infidelity by their mates—males resort to false alarms to interrupt copulation between their egg-laying partners and other males.[11]

In Formosan squirrels, females can mate with several males during their narrow window of estrus. Males, to secure their fatherhood after mating with females, will make false alarm calls, alerting their neighbors to watch out for predators that in reality don't exist. While the neighbors hunker down and wait for the nonexistent threat to pass, the females miss a critical moment when they might be inseminated by other males in the vicinity.[12]

False alarms, although commonly used, are far from the only trick. Cheaters have many other ruses up their sleeves, especially when sex is at stake. One particularly devious ploy is found in the Atlantic molly, a small fish. In this species, experienced males keep information about the type of female they desire—their sense of fish "beauty"—to themselves. They only reveal their true preference when alone with a potential mate. If another male is around—especially a young and naïve one—they will mislead him by courting a female they dislike, presumably one they find bland or ugly. By using this ruse, experienced males can prevent naïve males from learning what beauty consists of in their world and therefore reduce the number of male rivals.[13]

But no other animals can outdo primates in the realm of trickery. For instance, gorillas and chimps conceal their emotional expressions by covering their face with their hands. Chimps are known to hide objects behind their backs before throwing them. (That's one reason they can pose a danger to enemies, including human visitors at zoos.) Also, when subordinate chimps have an erection, they turn their backs toward dominant males.[14] Though we don't know whether they feel embarrassed, it obviously doesn't look good when your boss sees you sexually aroused. In the chimp world, this means that your presence poses a threat to the reproductive welfare of higher-ranking males, as you are now their rival in mate competition. Chimps are clearly aware of what each other think, an advanced cognitive ability known as a theory of mind.

Cheating can be used as a scare tactic for female self-defense against sexual harassment from undesirable males, particularly in animals where males are bigger and stronger than females. In these species, males often harass females to get them to mate. They may even force themselves on females against their will. Females in response resort to a variety of tactics to fend off unwanted advances. They may appeal to dominant males for help or stay near their preferred males for protection. But they can also resort to shrewd ruses for self-defense.

One of the quirkiest tricks females use for this purpose is found in mice. Some years ago, when my colleague Janxu Zhang and I were working on pheromone communication in mice, we were dumbfounded when we unexpectedly found two chemicals, 2,5-dimethylpyrazine and 4-hepatanone, in the urine of females. These chemicals are typical of ferrets, the archenemy of mice. How on earth did they end up in the pee of female mice?

Mulling over this surprising discovery, we came up with a hunch: female mice might produce the chemicals to mimic the presence of ferrets, and by this ruse, they could repel unwanted males. It was like an aerosol spray we use to keep pests away. We tested the idea by exposing male mice to the chemical compounds. Indeed, they quickly retreated, apparently in fear.[15] For quite some time after the research was completed, we were still in awe of this amazing evolutionary adaptation that enabled female mice to say, "No!" to their male partners. Because there is no polite way to say, "Not tonight, honey!" in the rodent world, you need something more persuasive.

In addition to trickeries for food and sex, many animals fake their ferocity to fight or their social status by bluffing. I am intimately familiar with this tactic from personal experience. Born in a small town in East China, I was gung ho for cricket fighting during my childhood in the 1970s. Every summer, my uncle took me to catch male crickets. We then pitted them in fights against crickets caught by other kids in the neighborhood. The contest took place between two cricket warriors in an arena that consisted of a tin can or a bamboo tube. Before the fight, we incited their aggressiveness by teasing them with a brush made of a grass stem. When they were fully aroused, we allowed them to fight. A winner emerged after about 5 to 10 seconds of biting and wrestling. The winner would chirp resoundingly while shaking his body back and forth in fits.

To increase my chances of winning a cricket contest, I faced the same challenge as someone betting on a horse race: I needed some way to spot a good cricket warrior *beforehand*. But how? Knowing that winners made loud sounds, I first picked loud chirpers. It didn't take long for me to realize that this method didn't work well, because fighting chirps were often used as a bluff. Cowards could easily fake their sounds as if they were fearless gladiators. Once they were in the ring, however, they quickly ran away—sometimes without even putting up a fight. So, chirping was an unreliable indicator of fighting ability. How about size, you may ask? Yes, it worked better. The problem was that all my rivals had the same idea. So, size didn't give me the upper hand.

After more experimentation, I eventually discovered a reliable predictor, a silent trait: the mandible gape—the size of the gap between the two mandibles. One summer, we were lucky enough to catch a cricket with an exceptionally wide mandible gape. Rather than bluffing, it always backed its battle chirps with a tenacious fight. As a result, we won all the tournaments in town that year.[16]

Bluffing in crustaceans—like in crabs and shrimps—is just as interesting as in crickets. Crustaceans grow by molting and are extremely vulnerable when they have shed their old armors and the new ones are still soft. Many crabs and mantis shrimps get around this vulnerability by bluffing, waving their menacing weapons at their opponents.[17] But they know they are paper tigers who can't engage in a real fight. As such, they will bail out quickly when challenged, despite appearing armed to the teeth. Unfortunately, fiddler crabs don't seem to be so wise. These crabs bear two major claws that vastly differ in size, with the small one for feeding and the larger one for fighting. If the fighting claw is lost, it will grow back. But for some reason, the regenerated claw fails to reach its original size. This doesn't appear to bother the crabs, for they will still use the large claw for bluffing as before. Only when a real fight breaks out do they realize that the poor replacement doesn't quite suit its purpose.[18]

Frogs and toads bluff, too, but what makes them stand out is their subtlety. They bluff by altering their voice. It's common knowledge that they croak to establish their territory and seduce females. Less well-known, however, is that physical conflict among these amphibians can be quite violent and taxing on their energy reserves. Body size is often the deciding factor for who wins a physical fight. For this reason, larger males are avoided by smaller males in some species. But how does a frog know the size of potential rivals, especially when he can't see them?

The answer lies in the pitch of calls when males advertise their territories. Larger males have larger larynxes and vocal cords, making their calls lower in pitch. The same acoustic principle applies to bulky subwoofers used in live performances by rock bands to amplify the effect of low-pitched sounds. (This applies to humans as well—larger individuals tend to have longer vocal folds and produce deeper sounds.) Unfortunately, voice pitch can be faked, so it is an unreliable indicator of size. Elizabeth Holmes, the CEO of the now defunct company Theranos, was said to fake a deep voice so that she would be perceived as powerful and dominant. (You can get a feel for her pretentious voice in public speeches available on YouTube.) I'm not sure whether she knew that she was employing an old trick in the playbook of many frogs and toads to deceive their rivals and cheat for mating opportunities.[19]

Dogs bluff by barking, to nobody's surprise. But they also declare their dominance silently by using their pee to signal their size. How can urine be used to cheat about size? We all know dogs are territorial. They mark their turf by spraying their urine on elevated objects such as tree trunks, fire hydrants, and power-line poles—a behavior you can't miss when you take a dog for a walk. Careful studies reveal that smaller male dogs tend to raise their legs higher than their larger peers when they pee. Thus, the height of their scent marks can be exaggerated. When other dogs in the neighborhood sniff the scent, they can be tricked into believing the dog who left his mark is larger than he really is.[20]

Many species of mammals also use urine or other body odor marked on prominent spots to claim their turfs, like leaving a note on the community bulletin board. Since the sources of their scents are mostly located in the rear part of the body, how can they advertise their impressive body size? One solution is to use an upside-down position by standing on their hands, so to speak. This acrobatic maneuver allows larger animals to post their scent marks higher up on large rocks or tree trunks.

Pandas, for example, use anal gland secretion and urine to mark their territories. Anal gland secretion is a sticky, pasty substance. It has to be squeezed directly onto an object like putting toothpaste onto a brush,

FIGURE 2.1. A male giant panda stands on its hands to mark its territory (photo credit: Rongping Wei).

limiting a panda's ability to cheat about its size. Urine, however, is different. When sprayed upward like a garden hose, it can go higher than the body. As might be expected, when pandas stand on their hands to mark territory, they almost always use urine—presumably to inflate their size (fig. 2.1).

Although bluffing is used by a wide variety of animals, it may not always work—especially when the test is a real fight. What other options do animals have when they are outmatched in a physical contest? One way is to recruit supporters. Indeed, some social animals can and do make an outsized fuss over a minor injury inflicted by one of their peers. It's a scenario familiar to soccer fans: after being fouled, a player may engage in theatrical displays, rolling on the ground while acting out unbearable pain, hoping the referee will rule in his favor.

Guess which animals would be most likely to use this trick? Monkeys and apes, of course.[21] For example, chimps, when at a disadvantage in a dispute, may scream hard and loud. We know they are cheating because they do so only when others are around. By overstating how badly they've been wronged, they can provoke sympathy and enlist support from bystanders.[22] In the same manner, juvenile Savannah baboons, when bullied by an adult, appeal to their mothers by screaming for help, a tactic often observed in human children as well.

🐾

Although animals contrive a wide range of schemes to deceive their peers, they more often cheat by a much subtler means: free riding—that is, putting less effort into a group project than the other members of the group. For instance, many young birds help their parents raise younger siblings, a behavior estimated to occur in 3–8% of bird species in the world. Helping allows young, inexperienced birds to learn critical survival skills, especially when they themselves are not yet ready to strike out on their own. Helping also benefits their fitness indirectly because they share, on average, half of their genes with any of the chicks in the nest. (This aspect of evolution is known as kin selection.) However, just as a half-full glass can be seen as half empty, so too are shared genes. A 50% genetic similarity can also be seen as a 50% genetic difference. This is why there is conflict as well as cooperation between the helper birds and their siblings.

For this reason, helpers may not pour their full effort into this communal reproductive enterprise. In carrion crows, for example, as many as 27% of helpers are sluggish workers, contributing less effort than they can toward helping raise their siblings. But if the dominant breeders are removed, these lackadaisical helpers will immediately increase their efforts in feeding the chicks. Clearly, they can be quite competent in food provisioning when they have to. They are by no means born lazy; they simply dodge their duty when they can get away with it.[23]

Among the giant babax, an Old World songbird the size of a large thrasher, 70% of chick food is brought by helpers. To encourage helpers to do their job, chicks, when their bills are touched by food, will release a fecal sac on the other end, which is then eaten by the feeder. So, the chick's body works like a pay-for-service exchange machine: goods in on one end and payments out the other. Though it's not entirely clear why chick feces are so highly prized, helpers are incentivized to feed the chicks for the reward. Cheating helpers, however, act as if they were feeding the chicks, though no food is actually delivered. But are they excluded from the food-for-feces trade? No. They've found a way out: they either put garbage in chicks' mouths, cheating to get the reward, or simply steal the feces from honest helpers when they put food into the chicks' mouths.[24]

Free riding is a common cheating tactic when an animal group faces outside aggression. The equivalents of draft dodgers are not rare among social animals. They've been identified among dogs, wolves, lions, lemurs, and monkeys. A pride of lions, for example, is organized around female kin, who defend a communal territory. Within the pride, however, individuals vary in their contribution to territorial defense. Some are good fighters, always leading the effort to protect their turf whenever facing a challenge from outsiders. Others are lazy, only stepping up when their help is essential. They contribute less than they should to defending the pride's territory. These free riders are nevertheless better in comparison with a few true slackers and cowards, who never offer any help even when it's desperately needed.[25]

Free riding can also be found in reproduction. This is best illustrated in social insects such as ants, bees, and wasps. In many cases, the queen is the only female in the colony to reproduce. All other females are workers that take care of colony chores. The queen monopolizes the reproductive rights for the entire colony and maintains her dictatorship by emitting a pheromone that prevents workers' ovaries from producing eggs. That is, the workers are reproductively suppressed. But unlike surgeries we use to neuter our pets, the chemical neutering process administered by the queen can be reversed. If the queen dies, some workers (called false queens) will quickly take over the regal right to reproduce. Workers in some species, apparently dissatisfied with their inferior role, may lay eggs clandestinely—even when the queen is still alive and well. (Since only fertilized eggs develop into females in these insects, unfertilized eggs laid by workers will develop into males.[26])

Such selfish behavior on the part of workers can compromise the efficiency of colony growth. Many social insects turn to policing to crack down on reproductive cheating by workers. When workers on patrol duty find a colony mate illicitly laying eggs, they either attack the cheater or destroy her eggs.[27]

Cheating by free riding and strategies to counter it are also seen in a wide range of species, from cichlid fishes and fairy-wrens to naked mole-rats. In these species, low-ranking individuals are often punished for being lazy.[28] In macaques, a subordinate member of the group can be punished by dominant individuals if it fails to make a "food call" after it discovers valuable food.[29]

𝔁

For most male animals, few things in their lives are more important than mating with females. To increase their mating opportunities, males

FIGURE 2.2. A male cichlid; note the egg-like spots on the anal fin (photo credit: Lixing Sun).

employ an awesome diversity of cheating strategies. For the rest of the chapter, we will explore this issue.

One example is an African cichlid (*Astatotilapia burtoni*) studied in my lab, a fish species with females that are dedicated mothers. They use their mouths as nursery chambers to provide extra protection for their eggs (often, some are unfertilized) and their young (called fry). If some of their eggs or fry accidentally escape from their mouths, they will quickly pick them back up. Males exploit this behavior as an opportunity to cheat. Adult males have evolved a roll of egg-like spots on their anal fins, which serve as a decoy that can cause females nearby to try to pick up the false eggs (fig. 2.2). A male then seizes this opportunity to release sperm, which may end up fertilizing the unfertilized eggs inside the female's mouth.[30]

This cheating behavior found in mouth-brooding cichlids seems dull when compared with other species. To ensure their cheating success, some males may throw everything—behavior, physiology, morphology—into the pot. In several species of salmon, for example, males can assume different sizes, each adapted to a specific mating strategy. In Chinook salmon, as well as Atlantic salmon, males come in three morphs. The normal type is known as the anadromous form. It migrates to the sea (from freshwater streams) and grows to adult size after four or five years. They return, mostly to the streams where they were born, to mate with females and then die soon afterward. Some males, known as jacks, take an easier way out. They come back from the sea to freshwater to reproduce a year or two earlier. Since they don't have enough time to fatten themselves fully in the sea, they are considerably smaller and less physically competitive. Jacks, therefore, are free riders in the sense that they take a shortcut in their life history.

But the champion cheaters among salmon belong to precocious males, also known as parr or precocial. These males spend their entire lives in streams, never going to the sea. Because they skip the rigors of migration, they avoid the radical change in body morphology and physiology needed to prepare for life in saltwater, a process known as smoltification.[31] They become sexually mature in their first or second year of life. With most of the material and energy funneled to growing reproductive organs, their bodies are tiny—only a few inches long. So, they are no match for the normal males, not even jacks, in a physical contest for females. They may even be attacked or eaten by larger males. But they make up for their physical weakness by cheating. During the mating season, they hide under rocks or wait in shallow areas that are inaccessible to larger rivals. When they see a female release eggs, they dash to the spot and spray their sperm onto the eggs. This unspectacular behavior can get the job done for these sneaky fish midgets.

A quirkier case of cheating for sex is found in a toadfish known as the plainfin midshipman, common along the West Coast from Vancouver to Northern California. I once took my kids on a crabbing trip in an area of Puget Sound near Seattle, but instead of crabs, we found toadfish in our traps. Most adult males, called type I males, have sound-producing organs modified from their air bladders. During the mating season in late spring and summer, they emit low-pitched hums to court females during the night. Because of this, they are sometimes called "singing fish" or "canary bird fish." When a large number of males are in an amorous mood at the same time, their monotonous hums can be loud and irritating. This phenomenon in the San Francisco Bay area in the 1970s led to conspiracy theories at first. Some thought the noise was from secret government operations, whereas others believed it came from factories illegally discharging pollutants in the night.

Although these sounds may be annoying for people, they're irresistible for female fish, who respond in droves upon hearing them. After locating a singing male, a female will release up to 200 eggs in the male's nest, built under rocks in the intertidal zone. She says goodbye to her eggs once they're laid and fertilized, leaving them in the care of the male. The male will try to lure additional females on succeeding nights until he gets thousands of eggs in his nest.

But the story of the plainfin midshipman doesn't end there, with the type I male happily tending his fertilized eggs. There are also type II males who neither sing nor build nests. Instead of expending energy on these laborious fatherly responsibilities, they save it all for growing sex organs—fast. They reach sexual maturity when they are only half the length and one-eighth the body mass of type I males. But they make up for their deficit in body size with extraordinarily large reproductive organs. Their testes can be fourteen times larger in size than those of type I males, relative to their body mass (color plate 3). Although they are eager to reproduce, they are unattractive to females, so type II males have to resort to trickery to find opportunities to mate. They do so by pretending to be females in color, size, and behavior. In this way, they gain access to type I males' nests without being noticed. Alternatively, they stay put near the entrance of a type I male's nest and fan their sperm into the nursery chamber, hoping some will hit the jackpot and fertilize the eggs.[32]

You may be wondering how males of the same species diverge in their tactical approach toward reproduction? The story is complex, but one enzyme, aromatase, plays a vital role. This enzyme can convert testosterone into estrogen, and thus affect sexual development by tweaking the testosterone-to-estrogen ratio in the brains of young fish. During sexual development in males, a lower aromatase level will lead to a higher testosterone level, turning a male into a singer, that is, a type I male. If the aromatase level is high, however, more testosterone is converted into estrogen, causing a male to become feminized and develop into a type II male who cheats.[33]

Sneaky kinds of cheating strategies are equally impressive in terrestrial animals. In some populations of the side-blotched lizard, for example, males come in three morphs, representing three different reproductive tactics. Orange males have an orange color on their throats. They are macho, aggressive lizards with high testosterone levels. They control large territories that include many females. The second type, blue males, guard smaller territories with fewer females. Therefore, unlike orange males, the blue males can slash down their defense budget as they don't need to spend so much time and energy fighting off other males. Since they are less aggressive, they have lower maintenance costs as well as lower levels of testosterone. As high levels of testosterone can compromise the immune system, blue males have a fitness advantage over orange males in this respect.

However, there is also a third type—the yellow she-male. Weak and homeless without territory, yellow side-blotched male lizards pretend to be females in appearance and behavior to fool the orange males in particular (see color plate 4). Once a yellow male gains a foothold in an orange male's territory, he does not behave like a grateful guest. Instead, he sneakily searches for mating opportunities under the nose of his ignorant host.[34]

The flat head lizard practices a similar sort of sneaky behavior. In this species, a small number of males turn themselves into she-males, like their counterparts in the side-blotched lizard. There is, however, a major flaw in their deception. Although a she-male can look and act like a female, his body odor can betray his true sexual identity. Because of this flaw in their disguise, she-males deliberately avoid close contact with he-males so that they won't be found out.[35]

This caution becomes unnecessary for cheating male red-sided garter snakes. They have evolved the capacity to make themselves smell like females. Garter snakes are probably the most common snakes in northern North America. You may see them in the forest, on farmland, and even in your garden. They are active from the spring to the fall but spend the winter hibernating communally in dens. In Canada, some dens can host as many as 10,000 snakes. When they're awakened by the warmth of spring weather, males emerge first. They come out en masse and wait near the den for females to appear.

How do males recognize females? They simply flick their forked tongues to pick up odorant molecules and examine them with a sensor called the vomeronasal organ (VNO for short). This enables males to sniff out females by the pheromones—methyl ketones—on their skin. If males sense squalene, a male pheromone, courtship comes to a halt.[36]

Unlike males, females emerge from the den alone or in small groups. But once a female appears, she's quickly surrounded by an abundance of amorous males, all vying for the opportunity to copulate with her. If you are ever in the right place at the right time, you may come upon a large number of snakes entwined together in a ball as if in a mating orgy. In reality, however, it's a one-against-all and all-against-one wrestling showdown among males. In the mating ball, there is only one female and 10 to 100 males, and in some extreme cases, up to 5,000 males.

With so many males so eager to mate, it doesn't take long for a female to finish mating. In fact, it could take as little as 30 minutes after she emerges from her den. Yet, the competition among males doesn't end with the conclusion of mating. The lucky male will ensure his paternity by sealing off the female's genitalia with a jelly plug, a physical barrier that makes it harder for other males to mate with her. In addition, he anoints the female with the male pheromone squalene to turn off other males' interest in pursuing her.[37]

Put yourself in the shoes of a male garter snake in a mating ball: what should you do? If you're a strong snake, you'll put forth all your effort and fight tooth and nail to win the physical contest. Even so, your chances are still slim because only one male will come out as the winner, and there is no consolation prize for second place. Your odds are even lower if you're not among the strongest in the bunch. What can you do to even them out?

You may have already gotten a hint from the side-blotched and flat head lizards. If you've guessed that the solution is to disguise yourself as a female, you're right. Like the lizards, garter snake she-males act like females to deceive the stronger males in the mating ball. Moreover, besides their pretense of female behavior, she-males smell like females as well. Their bodies are devoid of the male pheromone squalene.[38] These impostor tactics reduce the competition from other males and increase a she-male's chance to gain access to the prized female in the middle of the mating ball (fig. 2.3).

The cases I've cited represent only a few examples in vertebrates where cheating males coexist with honest males in reproductive competition. Such deceptive mating tactics have been found in a diverse

FIGURE 2.3. Garter snakes in a mating ball (photo credit: Oregon State University with a CC BY-SA 2.0 license, no modification made).

range of animals including insects, fishes, amphibians, reptiles, birds, and mammals. In fishes alone, males in 140 species have been observed using some form of deception to reproduce,[39] collectively called "alternative reproductive tactics." Deceptive tactics are called "alternative" because they are used by only a small minority of males but are not the main strategy used by the majority in a population.

Despite the diversity of specific methods, there is a high degree of commonality across species in males who deploy sneaking strategies to seek mates. The males are often smaller, unattractive, and physically less competitive.[40] They tend to mature earlier by channeling the bulk of their material and energy into reproduction rather than developing body morphs to impress females. Thus, cheating through deception and sneaking behaviors appears to be the result of males making the best of a bad situation in the mating game. To succeed in cheating, they look, smell, or act like females. Finally, sneaking males are only a minority, typically making up less than 5% of the population.

🐿

You may notice that some cheating schemes, such as false alarms, bluffing, and free riding, are used by nearly all animals, whereas others, such as alternative mating tactics, are adopted by only a small fraction of a population. Why is that the case? The answer is that cheating has both costs and benefits, which ultimately comes down to losses or gains in Darwinian fitness. (In research, fitness is often estimated by its proxies—time, energy, and risk—when it's hard to measure it directly.)

For cheating to be an adaptive strategy in an evolutionary sense, it has to provide a net benefit—that is, the benefits must outweigh the costs. Moreover, as cheating becomes more common in a population, its net benefit will decrease. When the net benefit of cheating dips below that of acting honestly, cheating will become maladaptive, no longer a viable strategy to thrive in evolution.

Use your hypothetical widget company once again for illustration. Assume everything else stays the same, but let's raise the rate of cheating—to the level that the risk of you being conned exceeds 50%. With that level of risk, you have to reject the proposed joint venture. Otherwise, the downside will outweigh the upside for your company's productivity. This commonsense case demonstrates a key point about the prevalence of cheating: it depends on the benefit-to-cost ratio for both cheaters and their victims. (Animals, of course, don't perform a conscious calculation. Evolution has already done the math by culling deviants who stray from the right course.)

Consider this from the perspective of the animal examples we've seen. False alarms, for instance, generally cost little to produce and have a high rate of return—such as food in crows, or paternity in barn swallows and Formosan squirrels. This creates a favorable benefit-to-cost ratio for cheaters. In addition, failure to respond to an honest alarm call carries a potentially fatal downside—being eaten by a predator. Since only cheaters know whether the message is true, receivers have no choice but to respond—just *in case* it's true. Therefore, the information warfare is asymmetrical, favoring cheaters.[41] Consequently, even frequent liars can get away with their tricks. That's why false alarms are so commonly used by so many animals for a wide variety of cheats.

Given the high benefit-to-cost ratio and information asymmetry favoring the cheaters, why haven't false alarms become even more ubiquitous in the animal world? One key factor that prevents cheating from getting out of hand lies in the response of the receiver. Being taken for a sucker is a painful experience in terms of fitness loss, and if victimized too often, receivers can fight back by sharpening their cognitive systems and exercising greater discretion in their response to alarms.

Indeed, receivers in many species develop criteria for whom they should trust and adjust their responses accordingly. Besides monkeys and apes, marmots and ground squirrels are known to discriminate against false alarms. Alarm calls from young animals, for instance, are mostly ignored—they are treated like the neurotic Chicken Little.[42] This is not because the young are born liars, but because they tend to be skittish, ringing alarms when the danger is not real.[43] This demonstrates that when an animal is identified by its peers as a liar, its reputation takes a nosedive and its chances of social survival are hurt. The harm of cheating boomerangs. This cost lowers the payoff for cheating and so keeps cheating in check, preventing it from causing more damage to the animal society or even making the communication loop collapse. (These examples tell us that the more a cheater lies, the less it will gain from it. This is known as frequency-dependent selection, which we will see many times in the coming chapters.)

The same can be said for bluffing. Just as with false alarms, bluffing carries a low cost but has a high potential benefit. Usually, testing whether a bluff is true runs the risk of incurring a high cost with little chance of gain. How many of us would be willing to reach out our hand to test whether a snarling bulldog or wolverine is merely bluffing? Again, we see that asymmetry in information regarding aggression favors the bluffer over the bluffed. That's why bluffing is a widely used strategy, even among animals that are unlikely to harm others, such as the soft-shelled crustaceans that we discussed earlier in this chapter.

Likewise, free riding has a high benefit-to-cost ratio. It requires minimal effort, yet the return can be large. If you question that formula, just ask someone who dodged the draft during the war in Vietnam—someone who came from a rich family and paid off his physician to falsely diagnose a bone spur in his medical record. Moreover, trying to guard against free riding carries a high cost. As many couples know well, blaming your spouse for not doing their fair share of housework is often more costly (creates more conflict) than simply doing the dishes yourself. That's why free riding is common in animal societies.

The problem of free riding gets worse as group size scales up, and free riding can become nearly cost-free.[44] Put yourself in the shoes of a lioness. It's easier for you to shirk your duty in a group of fifteen than in a group of five. Indeed, as society expands, more and more members may withhold their efforts. This can eventually derail communal responsibilities or collective activities such as defending a territory or raiding a neighboring village, leading to the tragedy of the commons, a social dilemma in which a community as a whole suffers when its members act on their individual interests. To prevent free riding from spinning out of control, mechanisms such as policing against cheating and other rule violations are increasingly necessary in larger groups. Have you ever wondered why elaborate laws and police forces are indispensable in modern societies, whereas in tribal societies, a few elders and some traditional social norms will suffice?

Cheating for sex using alternative reproductive tactics is, however, different from bluffing, false alarms, and free riding. Although the potential benefit (successfully passing on one's genes) is large, the potential cost is significantly higher. To be successful, sneakers are forced, in gambling terms, to go for broke: they have to sacrifice their body size to maximally channel material and energy into growing large reproductive organs. They're forced to live in the shadows of others. They especially need to avoid physical confrontation with male rivals. For those that take the path of she-males, their bodies must go through a whole battery of morphological, physiological, and behavioral changes, including, for some, suppressing their male-specific body odor. These changes are too big to be easily reversed. Once they've gone down that path, there is no return.

Sneaking also reduces the fitness of other males (for the loss of paternity) and females (for mating with an undesirable male). Thus, when sneakers become more common in a population, selection favors those who discriminate against cheaters, which reduces the benefit of sneaking. For these reasons, the number of sneakers in a population is usually kept to a small minority.

In this chapter, we have taken a tour through a variety of strategies that animals use to cheat. Although the examples we've provided are limited, there is no shortage of wondrous ways that evolution has devised for animals to lie and deceive. The cases highlighted here are far from comprehensive, but they are sufficient for us to see some patterns.

First, despite the vast diversity in the ways animals cheat, they all aim at gaining resources that are critical to survival and reproduction, including food, sex, and social status, among others. Second, one main way that cheaters succeed is by manipulating information in communication—by altering signals to disguise the truth. Somewhat tongue-in-cheek, I've designated this as the First Law of Cheating. It is the expectation of honesty in communication that enables lying to work.

Also, as long as cooperation can bring greater benefits than going it alone, some degree of cheating can persist. This is why cheating is common in many different species of animals as an alternative but viable and vital strategic option. Its very existence points to an evolutionary balance: honest signal senders and tricky cheaters can be equally adaptive, a phenomenon termed behavioral polymorphism in evolutionary lingo.

Finally, communication channels are leaky, and none of them appears to be safe against hacking by cheaters. Although the cases we've examined in this chapter are primarily in visual, acoustic, and olfactory communication, there is no reason or evidence to believe that other, less well-known channels (such as tactic, electric, seismic, infrared, and ultraviolet) are safer. It's likely that there is no such thing in nature as safe communication in an absolute sense.

The above takeaways are quite sobering, but they shouldn't make us pessimistic, because identifying problems is often the first step to finding solutions. Before we discuss anti-cheating strategies, we need to explore the second method organisms use to cheat.

CHAPTER 3

Nature's Eavesdroppers, Impostors, and Con Artists

In the late 1980s, I was conducting research on the water deer, a small, antlerless deer species the size of a large coyote. I stationed myself in a poor village in East China, where folks ate whatever they could get—deer, wild cats, raccoon dogs, geese, ducks, and even sparrows—for extra protein. Because many animals in the region were becoming rare or endangered, I tried to dissuade them from doing so. The good-natured villagers would normally comply—except when they wouldn't. One day that happened in a notable way.

Around dusk, Lao, an 18-year-old lad who was my occasional field assistant, spotted a mother mallard trailed by six ducklings on a local lake. "Ha, I can catch them," he said, his eyes glistening with excitement.

Knowing he was after the ducklings, I tried to stop him. "There isn't much to eat," I said. But he didn't listen.

He dragged out a small boat and yelled at me: "Jump on, quick!" No sooner had I hopped on board than he pushed it off the shore and dashed toward his target.

It didn't take long to catch up with the duck family. Lao cornered them on the bank, and the ducklings scattered in panic and hid in the tall grass. We got out of the boat to search for them. I quickly found one but didn't alert Lao. Just at that moment, he shrieked, "Lixing, the duck can't fly—she's hurt!" I turned my head and saw the duck on the ground, flapping her wings, seemingly in great distress. Lao turned his eyes to the duck. Obviously, a full-grown duck has a lot more meat than several ducklings.

The mother duck limped and staggered, struggling to get away, but Lao was right behind her, and her fate appeared to be sealed. Just when Lao was about to grab her, however, she took off suddenly with a light quack. She swooshed low as she passed above us with ease and grace.

When Lao returned empty-handed, the ducklings were nowhere to be seen. He kicked around in the grass where they'd first hidden, but nothing was flushed out. As darkness drew near, we had to call it a day.

"If you don't know their tricks," I remarked wryly, while chuckling to myself, "you won't get any wild duck on your dinner table." The heroic mother had saved her children.

Like the brave mother mallard, lots of animals fake injuries to divert a predator's attention. Biologists aptly call such ruses distraction displays. Plovers, shorebirds that include the familiar killdeer commonly seen on North American lawns, are probably the most familiar of such animal tricksters. On one occasion, a killdeer faked injury when I accidentally stepped too close to her nest in a schoolyard. She couldn't know I wouldn't take the bait. On the contrary, her deceptive ruse alerted me to search in the opposite direction. I quickly located her nest on a sandy patch, which was a training ground for the shot put! Thankfully for her, it was summer, and school was out.

If you've ever lived on a farm, you may have noticed that barnyard hens, when chased, may suddenly freeze. If you assume they're giving up struggle and accepting the inevitable, you'd be wrong. It's actually a trick that sometimes works to their advantage. Struggling can arouse the killing instinct in some predators, such as cats. The house cat, for example, is known for "playing" with—molesting, to be more precise— its prey, such as a mouse, until the victim becomes motionless. Often, at that point, the cat loses interest. It may ease its grip on the victim, or even wander off in search of something more exciting. This, however, gives the mouse an opportunity to escape its almost guaranteed death.[1]

Feigning death is not uncommon, it turns out. Many species of lizards "play possum" when approached by snakes, which are unattracted to prey that appears dead. Woodchucks, also known as groundhogs, freeze when they perceive you can outrun them and cut off their paths to their dens—as I once tried to do some years ago. Antelopes use the same trick when they're captured by cheetahs or leopards. But if the predators loosen their grip prematurely, the victim can suddenly come back to life and run away. This scenario is popular in many wildlife programs.

Feigning death may save your own life when pursued by a large predator such as a grizzly bear, as you may already be aware. Indeed, without a can of bear spray or any effective weapon, your best bet is to play possum, hoping the grizzly will leave you alone. Grizzlies are territorial and tend to attack people when they feel threatened. So, if you lie down motionless, they may no longer see you as a potentially dangerous intruder. If you're lucky, they'll curtail their aggressive pursuit. The bad news is that this tactic is far from foolproof—not all predators are easily tricked, particularly when their stomachs are empty. And you should never try this when facing a black bear, a species that *does* eat dead meat. Feigning death may only make you more attractive as a food source for this type of predator.

Early European settlers in America understood this tendency in predators and used it to protect their interests. In some areas, large predators such as wolves and cougars preyed on livestock, taking a serious toll, but people found a workaround in a breed of goat that would "faint" when stressed or startled. Some ranchers bred "fainting" goats so that when their livestock was attacked, the unfortunate goats served as an easy target to stop the predators from going after more valuable animals, such as cattle and sheep.

The fact that you can breed fainting goats is evidence of a genetic basis for fainting behavior (meaning, a male fainting goat bred with a female fainting goat can produce fainting offspring). The tendency to faint when stressed is formally known as congenital myotonia. It is due to a recessive gene found in goats and other animals, including humans. The muscles of the animals with two copies of the gene (called homozygous recessive) will seize up and cease to function at a critical "fight-or-flight" moment. As a result, they fall on the spot as if they were dead.

Faking injury or feigning death are just two examples of a general type of cheating where one animal takes advantage of the loopholes in the cognitive system of another. This is what I've designated the *Second Law of Cheating*, which is the biological foundation of deception. Unlike the First Law, where cheaters alter the meanings of honest messages in communication, the Second Law involves exploitation of biases, weaknesses, or deficits in the cognitive system of another animal. Although the two examples we've cited so far involve one animal deceiving another of a different species, it occurs among peers within the same species as well.

In this chapter, we'll meet some of nature's most fascinating impostors to gain an appreciation for the creative ways animals deceive others. We'll examine a diverse assortment of cases that illustrate how organisms practice the Second Law of Cheating, that is, deception. We'll explore ways that unfriendly interactions between the cheating and the cheated can lead to a dazzling array of morphological, physiological, and behavioral adaptations from camouflage and bluffing to many forms of mimicry.

𝄐

Before exploring cheating between different species, we need to understand why loopholes in sensory and cognitive processes exist. Let's begin with a field trip to Mexico.

A species of blind fish known as the tetra lives deep in caves in central and eastern Mexico. Its ancestors had functional eyes, but the tetras gradually lost their eyesight as they adapted to life trapped in lightless caves. Wouldn't it be nice to have eyes, you may think? But what are eyes for? In the absolute darkness of a cave, eyes are not only useless but also wasteful, as they have a real cost in material and energy (for the neurons and neural wiring) to stay functional. Additionally, fish eyes are a primary portal for many pathogens and parasites—just as windows in a house allow occupants to see out, they also enable robbers to break in.

When eyes contribute nothing yet hurt the survival and reproduction of the fish, they become a major burden. Disfavored by natural selection, the eyes lost the ability to fight the buildup of harmful mutations that crop up from time to time. This eventually spelled doom for vision in cave-dwelling fish. By and large, this is how one rule of natural selection—use it or lose it—works.

Losing eyesight actually is good for the blind fish. It allows them to use the energy that was normally expended on developing and maintaining eyes on more vital body parts and activities. In the cutthroat world of an evolutionary race, every competitive edge counts. As a result, the blind fish thrived, whereas those that indulged in the useless luxury of sight died out.

The evolution of the blind tetra provides a moral of sorts: you can't be good at everything. This is particularly true for sensory systems. With limited resources, animals must prioritize resource allocation. For this reason, it's impossible for an animal to evolve the sharpest eyes, ears, and nose all at the same time.[2] The situation can be compared to government budgeting. If you want to spend more money on defense, you have to cut back expenses elsewhere such as farm subsidies, welfare benefits, and other programs. Were the financial resource infinite, it would be unnecessary to debate how money should be spent.

But there is one key difference. Government budgetary decisions are often made for political reasons, not necessarily for efficiency. Evolution, however, always operates on the principle of efficiency. Otherwise, organisms would be outcompeted by their rivals. This is why natural selection favors senses that are most beneficial for fitness, while letting trivial luxuries—like eyesight in a totally dark cave—degrade, decay, and eventually disappear over time.

We humans have good vision, comparable with most mammals, though not as good as most birds. Our hearing is fair, but our olfaction (smell) is gravely inferior to that of many mammals such as dogs, pigs, or rats. Worse, we are nearly completely deficient in our ability to sense infrared, ultraviolet, ultrasonic, and electric cues and signals in our environment, whereas many other animals can. For instance, vipers can sense infrared wavelengths; many birds can see ultraviolet light; rodents can hear ultrasonic sounds; and electric eels can detect electrical cues. Also, we are helpless at such things as using echolocation to find objects as bats do, or sensing seismic vibrations from miles away as elephants do.[3] All of these relative weaknesses are potential sensory

FIGURE 3.1. A moth (*Phalera takasagoensis*) mimics a dead tree branch (photo credit: Jingang Li).

loopholes that make us vulnerable to exploitation by other species, just as Lao was duped by a mama mallard duck. Luckily, modern humans can make up for these flaws and deficiencies by relying on scientific instruments. But other animals can't. So, their sensory loopholes are wide open to exploitation—they almost invite it. And this is why cheaters can get away with their trickery by applying the Second Law of Cheating.

The term "sensory exploitation" was first proposed by evolutionary biologist Mike Ryan in the late 1980s. It's often used in the context of sexual selection, which we'll elaborate on in chapter 5. Here we'll look at the idea more generally as it applies to all situations where one species takes advantage of the biases, weaknesses, or deficiencies in the sensory system of another species. These biases, weaknesses, and deficiencies are collectively called *cognitive loopholes* in this book.[4]

It is unsurprising that the broad occurrence of cognitive loopholes offers a major opportunity for others to exploit, leading to the evolutionary invention of many types of mimicry, where one species can make itself look like another species or a dead object (fig. 3.1).[5] For example, guppies, especially female guppies, are attracted to orange color. Prawns, which prey on guppies, have evolved orange spots on their pincers to attract guppies, making it easy for the prawns to catch the fish. Death adders, a deadly Austrian snake, wriggle the tips of their tails like live worms to lure in their prey lizards, and bolas spiders imitate moth sex pheromones to reel in moths.[6]

These are only a few examples, but they illustrate how the sensory preferences of prey can be exploited by their predators. Prey species, too, have evolved a wide arsenal of tactics to fight back, including exploiting the cognitive loopholes of their predators. In the following sections, we'll see how animals use the Second Law of Cheating to scare, confuse, distract, fool, or mislead other species, all by taking advantage of cognitive loopholes in their victims.

𝕬

In our tour through the captivating world of animal impostors, hustlers, and con artists, let's start at Englehorn Pond in Ellensburg, Washington. Small, shallow, and swampy, the pond is unexceptional. What makes it stand out is that it's the home of hundreds of tiny Pacific treefrogs. These frogs come primarily in two distinct color morphs: gray and green (see color plate 5). Some locals call the gray morph "wood frog" and the green morph "treefrog," as if they were two different species.

When James Stegen, Cory Straub, Genevieve Phillips, Chris Gienger, and I observed these frogs in their natural habitat, we found a near-perfect match between the color of frog and the color of the background during the day: the green morph hangs out on green cattail leaves and the gray morph hangs out on mottled-gray leaf litter. Since frogs on the "wrong" background could be easily spotted and picked off by hungry birds, they have evolved some mechanism to change their appearance and thus exploit a visual weakness of the birds. We wanted to know what the mechanism was that enables frogs to match their background. Mulling over the possibilities, we came up with two hunches: either the frogs know their own color and select a matching background for where to spend the day, or they change their own color to match the background where they find themselves. We called the first possibility "know-your-own-color hypothesis" and the second, "chameleon hypothesis." We decided to test which was correct.

We took green and gray frogs to the lab, put both kinds in Petri dishes with either green or gray background and took digital photos. We then used Adobe Photoshop to analyze how frogs blend themselves into the background over several hours. After some data crunching, the verdict was in: both green and gray frogs can modify their body colors to match their backgrounds slightly better. But their capacity in short-term color change is quite limited. They can't switch freely between the green and gray morphs in a matter of hours or even days.[7] Therefore, the chameleon hypothesis was out. Apparently, they do know their own color and use this information to find a fitting spot to spend the day.

But even with the support of our lab results, we still felt unsure as to whether this know-your-own-color hypothesis indeed works in nature. So, we took our experiment to the field, again at Englehorn Pond. The frogs spent the day on land plants, fattening themselves by ambushing tasty bugs, but they conducted their most important business— reproduction—in the night, when males give off loud croaks on pond weeds, broadcasting their carnal intent to females. Since the frogs could move everywhere between the pond weeds and the land vegetation on a daily basis, how could we find and track individual frogs for weeks or months? We soon found, to our delight, that this concern had been totally unnecessary.

As it turned out, tracking individual frogs was quite easy when we discovered that our little "research participants," a term preferred by psychologists, have a remarkable level of site fidelity. Many of them returned to the same spots day after day, as if they were helping our research. On many occasions, we found that the same frogs perched on exactly the same leaf blades in a patch of cattail plants. It's still a mystery as to how the tiny frog can navigate so accurately in such a convoluted habitat. If we were treefrogs, we would need a high-precision GPS system to locate the same spot day in and day out.

With the "help" of the frogs, we were able to track dozens of them individually for months. In the end, while we validated our results obtained in the lab, there was also a minor surprise: although most frogs can only adjust their color slightly in the short term, a few of them did succeed in making the switch between the two color morphs over several months.[8] Biology is indeed a science of exceptions.

ℱ

Many animals are far more versatile than these treefrogs in changing their appearance and, thus, far better in using the Second Law of Cheating. Chameleons, octopuses, and stick insects are among nature's most spectacular impostors. The ability to change color and shape to imitate other species or blend into their background is known as Batesian mimicry, credited to the British naturalist Henry W. Bates, the first to notice striking visual similarities between toxic and nontoxic butterflies.

A dedicated Victorian naturalist, Bates ventured into the jungles of the Amazon in 1848 with Alfred Russel Wallace, the codiscoverer with Darwin of evolution by natural selection. Wallace headed back to England four years later with thousands of specimens. Unfortunately, the specimens were all lost when his ship, the *Brig Helen*, caught fire and was abandoned. The passengers drifted in two lifeboats in the Atlantic Ocean for ten days before being picked up by a passing cargo ship. The ordeal left Wallace's body "scorched by the sun, hands, nose and ears being completely skinned," he wrote in a letter to a friend. Luckily, Bates wasn't on the ship. He stayed on in the Amazon until 1859 and sent home specimens of more than 8,000 species that were unknown to science up until then.

While in the jungle, Bates was attracted to the exquisite *Heliconius* butterflies. He was intrigued that several nontoxic species of butterflies appear and behave like the deadly *Heliconius*. Even though he was one of the world's foremost experts on butterflies, he could still be duped by these impostors. Despite some hunches about this uncanny resemblance, he couldn't come up with a good explanation. After he returned home, the puzzle lingered in his mind—until he was struck with an epiphany while reading Darwin's new book, *On the Origin of Species*: the butterfly mimicry he'd observed could be the work of natural selection.

If a nontoxic species could mimic a toxic species, it would deter predators and increase its chances of survival. He published the idea in 1862. After reading it, Darwin was so elated that he wrote to Bates, telling him that his paper "is one of the most remarkable & admirable papers I ever read in my life." These exuberant words were noticeably out of character for Darwin, who was known for being modest, understated, and habitually reserved.

Most animal mimicry we encounter is Batesian in nature: an otherwise edible or harmless species masquerading as a noxious or poisonous species, or even as inanimate objects such as logs, leaves, rocks, or bark. This enables them to fool or to hide from their predators or prey. Here is the colorful way Wallace described this variety of evolutionary adaptation: "They [the mimics] appear like actors or masqueraders dressed up and painted for amusement, or like swindlers endeavoring to pass themselves off for well-known and respectable members of society."[9]

Wallace wasn't exaggerating. Cases of dramatic mimicry are everywhere in nature. For example, have you ever noticed that some flies, called hoverflies, resemble bees or wasps (see fig. 3.2 for an example)? They look scary, don't they? In fact, there are 6,000 species of hoverflies in the world that gain protection by disguising themselves as something much more menacing. For many of them, it's more than simply a visual trick—they even buzz like bees and wasps to complete the illusion.[10] If a predator applies the same logic many humans use—"If it walks like a duck and quacks like a duck, it's a duck."—it will be deceived.

False eyespots found in many insects are another ploy widely believed to deter predation by birds.[11] This has been rigorously tested in butterflies and moths that bear eyespots on their wings.[12] By the same token, some animals are known to suddenly flash bright colors or emit loud sounds to scare away predators, a tactic known in biology as startle display.

The logic behind some deceptions may not be immediately obvious, and deeper analysis may be needed to understand. For example, pandas are known for the dramatically contrasting black and white patterns in their fur. How can such bold markings protect them? The answer: disruptive camouflage. Predators such as leopards and tigers have to

FIGURE 3.2. A hoverfly with fuzzy, yellow hair
on its body like a bee (photo credit: Lixing Sun).

recognize prey animals first before deciding whether to pursue them.
The broken contour of a panda's fur makes it hard for the predators to
form an image of a fluffy, clumsy, and tasty animal, especially against the
snowy background common in the panda's natural habitat, as a recent
study shows (fig. 3.3). Like a connect-the-dots puzzle, where we don't
recognize the image until enough dots have been joined, disruptive
camouflage works in the opposite way by removing the connected dots,
making the image harder to identify.

Zebras are widely believed to be a version of visual trickery called
motion dazzle—also known as dazzle camouflage or razzle dazzle—to
confuse lions, their main predator. However, evidence in support of this
hypothesis isn't robust. Recent studies show that the dazzle effect may
not be primarily aimed at lions but at something far less impressive:
irritating blood-sucking flies, especially tsetse flies. Apparently, flies
don't like zebra stripes, which interfere with their landing.[13] A group of

FIGURE 3.3. Disruptive camouflage in the giant panda in near (above) and distant (below) view. As the white color of the fur merges with the snowy background, the contour line of the body shape is broken, making it harder to detect (Nokelainen et al. 2021) (photo credit: Rongping Wei).

Japanese scientists found that painting cows with black-and-white stripes can cut down the number of fly bites by half compared with the unpainted control.[14] If this result is confirmed, we can expect to see more zebra-striped cows happily grazing in grasslands, much less bothered by flies.

In addition to zebras, a wide range of animals such as snakes, fishes, insects, and squids use motion dazzle to disguise themselves from pests and predators.[15] The same adaptation among a highly diverse set of species is known as convergent evolution—a more or less identical solution evolved to meet the same need. For example, eyes in squids and vertebrates are stunningly similar in structure, but they evolved independently with no common origin, quite like the multiple, independent inventions of the wheel in isolated ancient cultures.

In the case of motion dazzle, the visual illusion has evolved independently in many species that need protection against nasty bugs and hungry predators. You can experience the efficacy and confusion of motion dazzle yourself. (See the rotating snake illusion in color plate 6). My wife once had a blouse with fine black-and-white stripes. It made me feel dizzy every time I saw it at close range. After I complained a couple of times, she never wore it again. Interestingly, such a motion illusion may also work for fish and a vast range of visual animals.

Visual mimicry, especially camouflage, inspires creative expressions in art, fashion, and other areas of design. Camouflage is most often associated with military applications (fig. 3.4). Military camouflage was first conceived by British evolutionary biologist Edward B. Poulton and American artist Abbott H. Thayer in the late nineteenth century. Thayer got more credit for pitching the idea to the military during World War I. Although lacking scientific credentials, he persisted in the face of repeated rejection and ridicule. Camouflage was gradually adopted for military uniforms and equipment by American, British, and French forces. During World War II, camouflage was so broadly used by military forces that deploying troops in the battlefield without it was considered suicidal.[16]

Today, camouflage is essential in the military. Moreover, the arms race between camouflage and camouflage detection has migrated from our evolved cognitive systems to our evolved technologies. A whole

FIGURE 3.4. Motion dazzle inspired by zebra stripes: in T-shirt design (advertisement displayed on Walmart.com) and in military camouflage (British patrol gunboat *HMS Kildangan*) (photo credit: Imperial War Museums).

range of hi-tech ideas and gadgets have been explored and mobilized to minimize the detectability of military personnel and equipment. Cutting-edge technology can even make body heat invisible to infrared sensors. The United States still holds a significant military advantage over all other nations in this area, and their stealth aircrafts can evade even the most advanced radar system.

Batesian mimicry has two major limitations, however. One is that mimics have to be rare to be effective. Darwin noted this in 1859: "The mockers are almost invariably rare insects; the mocked in almost every case abounds in swarms."[17] If mimics are too common, their deception will be less successful. This is like deterring thieves by putting out a yard sign, falsely claiming your house is protected by the world's most advanced electronic security system. This works only if few in your neighborhood do as you do. If many signs aren't backed up by real alarm systems, they will act only as harmless scarecrows, and the deception will no longer work. Putting it in economic terms, the marginal profitability of lying goes down when you lie too much. That is, too many lies undo the very purpose of lying. So, you can't just talk the talk; you have to walk the walk.

(A quick aside: Decades ago, the advantage of being rare was often narrowly referred to as the rare-male effect. In a situation when males are the rarer sex, they can sire more offspring on average than females can.[18] Conversely, females can do better when they are the rarer sex. So the evolution of sex ratio will be like a pendulum, swinging around the stabilizing point of a 1:1. This is why, when we study sex ratio, we see an even split between males and females in most animals. In general, a fitness advantage declines as a trait becomes common. This is called frequency-dependent selection proportion, an important evolutionary concept that we referred to earlier and will see more later.)

What can mimics do when they become increasingly common, which hurts their chance of success? Fortunately, there is an evolutionary rule of thumb that applies: whenever there is a problem, usually there is also a solution. In a butterfly species in West Africa, for instance, the solution is five different types of eggs. When metamorphosing into adults, each of the five types can mimic one of the five toxic butterfly species living in the same area.[19] When mimics of one type become common and the efficacy of mimicry goes down, those that imitate a rarer type have a better survival rate. So the cycle goes on.

But multi-target mimicry in the African butterfly pales in comparison with that found in squids and cuttlefishes. Some of these marine animals have taken the Second Law of Cheating to the level of fine art. Not only can they randomly switch among several color morphs to mimic others in their community, but they can also execute their tricks at lightning speed. Sometimes, all it takes is one minute for them to switch among several color morphs. In one species of octopus, males are so good at masquerading as females that even females of their own species can sometimes be fooled.[20] Apparently, such a maneuver can prevent their visual predators from forming a reliable search image.

Another problem with Batesian mimicry is that naïve predators are often foolish enough to try new things. If you imagine this problem from the perspective of a prey animal whose body is protected by deadly toxins, you still risk being eaten, even if the naïve predator ends up dying too. What can you do to increase your odds of survival in this case?

One answer is to assume the appearance of other toxic prey. Or better yet, take on coloration that is broadly seen by predators as dangerous, if such exists. This would lessen your chances of death. Nature does indeed have a universal color code that indicates danger, most often bright yellow, orange, red, or blue (color plate 7). Since these are usually linked to extreme toxicity, they are known as warning colors or aposematic coloration, coined by none other than Edward Poulton. Many predatory animals are *innately* inclined to avoid them without having to learn by hard (and possibly fatal) experience.

In contrast to Batesian mimicry used for concealment or camouflage, warning colors stand out for the purpose of making their bearers conspicuous, like hunters or road workers wearing bright orange vests to warn other hunters or motorists. Animals use bright colors to make a bold statement, "Eat me and you die!" When different species of toxic animals converge on similar color patterns, it's known as Müllerian mimicry, named after the nineteenth-century German naturalist Fritz Müller, who used a mathematical model to show how this type of mimicry works.[21]

Though less common than Batesian mimicry, Müllerian mimicry in the natural world is not rare. Among the best-known cases are butterflies such as the viceroy and monarch, as well as many species of *Heliconius* (color plate 8). All noxious or toxic animals with similar warning colors in a region are likely examples of Müllerian mimicry.[22]

𝍊

Just in case you get the wrong impression that all between-species cheating is based on the Second Law, the First Law of Cheating is also applied in many such situations. This is especially true for eavesdroppers, where one species breaks into a communication system used by another.

Among such master codebreakers are fireflies. One example is a species of *Photinus* fireflies that flirts using blinking bioluminescence, like two ships communicating with flashing lights. Female fireflies live only about two weeks, and during that short period, they have to mate and produce about 100 eggs before dying.

The male often takes the lead in the courtship ritual. He flashes a few times and then waits for a response. If a female in the vicinity flashes back, the male answers, and responsive flashing continues as the male moves closer to his potential mate. During the flirting process, females are drawn to males with longer and faster flashes, indicating they'll be able to provide large, nutritious nuptial gifts—the fireflies' version of diamond rings.

Males are thus driven to blink as vigorously as they can, hoping to impress prospective mates. However, their flashes may also draw the unwanted attention of their deadly enemies, the females of another firefly species in the genus *Photuris* (not to be mistaken with the victim *Photinus*). These femme fatales can mimic the mating signals of *Photinus*. When unwitting *Photinus* males come for a date, they end up instead as dinner for hungry *Photuris* females.

The *Photuris* predators not only get nutrients but also a dose of protective chemicals, lucibufagins, from their *Photinus* prey. Lucibufagins are toxins that make fireflies distasteful for birds and other animals. These toxins are precious because they can protect the *Photuris* from their own predators—birds and spiders—higher up on the food chain. Since *Photuris* can't produce these chemicals on their own, they depend on *Photinus* meals. It's a great deal for the deceivers, as they gain both nutrition and protection at the same time.

A similar game of visual deception takes place in the deep sea, home to around 320 species of anglerfishes, also known as frogfishes, monkfishes, and other varieties. Half of these fishes live 300 meters below the surface, where virtually no sunlight penetrates. In many fish species, the back fin is turned into spines. But in anglerfishes, it's shaped like a lure to entice their prey. Structured like a mini-fishing rod, the tip of the lure glimmers in the dark from the millions of glowing bacteria. That's how they catch their meals.

But this is only half the story. In some species, only females use this trick. Males are morphologically different, often only a fraction of the size of a female. It's hard to believe that the tiny males belong to the same species (fig. 3.5).

Males use females' glowing lure as a beacon to locate a potential mate. When a male finds a female, he grabs onto her. And then a chain of amazing events takes place: the male slowly fuses into the female's body, connecting his own body to her blood vessels for food and oxygen. This parasitic lifestyle is so lopsidedly beneficial for the male that

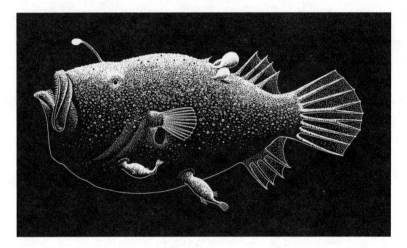

FIGURE 3.5. A large female and two tiny males in the triplewart seadevil, a species of anglerfish (copyright: Richard Ellis/Science Source).

he doesn't need to work on his own anymore. Consequently, his digestive system gradually dissolves, and the sensory system, including eyes, degenerates. All of the material and energy he saves is rerouted to his reproductive system, which quickly matures. He is ready to fertilize eggs released by the female on short notice. Because of their large size difference, several males may latch onto the same female.[23] Biologists call it a polyandrous mating system, in which one female mates with several males.

Leaving the deep sea, we move to the shallower, warmer part of the ocean, where we find unusual species called cleaner fishes. Cleaner fishes are famous for providing medical and dental services to other fishes by removing parasites and other assorted gunk that builds up in their clients' mouths. They are colored bright blue or yellow to signal their professional status. These color patterns are used to advertise their cleaning services, similar to a traditional barbershop marked with a spinning red-white-and-blue pole for identification. Business at some fish-cleaning stations can be so brisk that the clients have to line up, awaiting their turn. But as you might expect, the booming business can also attract unwanted visitors who come to take advantage.

One example is a deceptive fish called the bluestriped fangblenny that lives in the Pacific coral reefs. Fangblennies are good at changing body color and are skilled in mimicking the juveniles of several species of cleaner fish. They evolved this ability not to clean other fish but to specialize in sneaking up and biting chunks of flesh from fish clients coming in for a cleaning. For this reason, fangblennies are a drag on the cleaning business. When they show up at a cleaning station—like gangsters running a racket—the number of client fish can drop by as much as 40%.[24]

℣

It's wrong to assume that mimicry needs to be perfect, despite the many stunning cases we've examined. To exploit the cognitive loopholes of another species, you only need a good enough disguise to fool your target. Often a very crude mimic will suffice. This is why, for instance, blister beetle larvae—parasites on the nests of solitary bees—can succeed in duping their hosts.[25] They play a simple but effective trick. Hundreds of them swarm into a ball that is a crude analog—in size, color, or perching location—of a female host. They can enhance their mimicry by synchronizing their movements. Even so, it would be unlikely that a human would mistake a ball of beetle larvae for a female bee. For a poor-sighted male bee, however, the fake female is indistinguishable from the real thing. But when he tries to copulate with a beetle larvae ball, he ends up giving a free ride to the parasites, carrying them to their desired destination—the bee nest.[26]

As with the poor visual acuity found in solitary bees, all animals have limits to their cognitive capacities. For this reason, false patterns in mimics—that to us seem obviously bogus—can often work quite well. For instance, harmless king snakes mimic deadly coral snakes, as both have red, yellow, and black rings. The patterns, however, are arranged quite differently (color plate 9). Experiments show that the snakes' predators can't tell the difference.[27] Why? Perhaps because coral snakes are so deadly that their predators take a conservative precautionary stance—"better safe than sorry."

This isn't difficult to grasp. Imagine yourself as a predator, perhaps a hawk. From on high, you spot what appears to be a tasty snake. If you hesitate for even a few seconds, your meal might be gone. Time is of the essence, and you must act quickly. But there's a twist—a major one. If you mistake a coral snake for a king snake, it will cost your life. In the tradeoff between a meal and your life, the choice is obvious. For a predator, natural selection has shaped this decision-making process to favor the conservative choice—long-term benefits of survival over short-term benefits of a single meal. Thus, there is no need for the king snake to be a perfect mimic of the coral snake's colors to win this round of the evolutionary arms race.[28] The same logic applies when you're deciding whether or not to eat a wild mushroom. If you are not absolutely sure, don't let yourself fall prey to thoughts of culinary reward. The cost of your life would be too heavy a price to pay.

There may be another reason that the color order of the rings matters little for king snakes. Memorizing the sequence of ring colors appears to be beyond the capabilities of many birds. We know that many birds, including raptors, are quite smart. Crows, for example, make use of tools and solve simple puzzles; some are known to swoop at people to make them spill French fries out of their hands. Many birds can count and remember the number of eggs in their nests, for a good evolutionary reason: they don't want to lay too few, which would mean underinvestment in reproduction, or too many, which would mean overinvestment. Either of these miscalculations will reduce their fitness.

Memorizing the sequence of ring colors is not an effortless task, even for humans. (However, we can resort to mnemonic tricks like rhymes to facilitate identification. Here's one example: "Red touches yellow kills a fellow / Red touches black—venom lack / Yellow touches red, soon you'll be dead / Red touches black, friend of Jack.") So, king snakes are protected from predation even with the wrong order of colored rings.

𝔧

Because humans perceive the world predominantly by sight, acoustic mimicry seems relatively trivial. But that's not the case for many other species. Biologists have discovered numerous examples of using sound to fool others in a wide variety of animals and situations. And just like its visual counterpart, acoustic mimicry is based mainly on the Second Law of Cheating.

One of the most familiar varieties is bluffing by mimicking the sounds of a more menacing animal. This is extremely common in reptiles, birds, and mammals. Burrowing owls, for example, use holes in the ground to nest. When threatened, they hiss in ways that mimic rattlesnakes. This enables them to fend off a variety of predatory animals that otherwise might attack to get at their eggs and chicks.

Birds are best known for acoustic mimicry. Among the most familiar are mockingbirds, catbirds, and parrots. Some are so versatile that they can copy the sounds of woodcutting in sawmills and the shrieks of flying bombs, in addition to human language. They imitate these sounds for a variety of reasons—to attract females, to increase feeding efficiency, to alert their comrades, to facilitate social interactions, and to protect their territories.[29]

Lesser-known and fascinating cases of audio mimicry are found in insects. For example, if you pinch a sphinx caterpillar gently, it whistles. The whistling doesn't seem to be related to pain. They are simply hoping to drive away their predators—small songbirds like chickadees—by mimicking the chickadees' alarm calls. If you record caterpillar whistles and play them back, songbirds in the vicinity will dive for cover. In a similar manner, some toxic species of tiger moths emit a clicking sound to warn predatory bats that they're (truly) toxic, so the bats would be better off leaving them alone. The bats clearly get the message because they quickly learn to avoid the moths. As you might expect, such a ripe opportunity doesn't go unnoticed by potential impostors. Several harmless moth species also produce clicking sounds to mimic the acoustic aposematism.[30] (A pop quiz for you: What kind of mimicry is this, Batesian or Müllerian?)

🦋

The two laws of cheating can also be exploited through olfactory communication, communication through smell. A case in point is the Alcon blue butterfly illustrated in the book *Cheats and Deceits* by Martin

Stevens, a leading researcher in animal cheating. Before we get into the story of this butterfly, a question: given all the options the world has to offer, where would you choose to live if you were an insect larva and had to live in the home of another species?

One answer is an ant colony. There, you can get the best food delivered by nature's most diligent workers. You can also get maximal protection by nature's fiercest warriors, armed with biting jaws and nasty chemicals such as formic acid. Willing to sacrifice their lives, ants are formidable by dint of sheer numbers and their ability to act in unison like well-organized military troops. Given these advantages, as many as 2,000 species have evolved ways to mimic ants so they can cheat for protection and food.

Jumping spiders are among the deceptive impostors.[31] Some not only emulate the appearance and actions of individual ants but also copy their social behavior. One jumping spider by the name of *Myrmarachne melanotarsa* is normally solitary like most spiders. But when faced with predators such as birds, other spiders, and insects, dozens or hundreds of individual spiders can organize themselves into a faux army of ants, mounting a much more effective defense by working as a group.[32] To us, the fake is obvious because ants have six legs and spiders have eight. But it's convincing enough to fool predators that can't count the number of legs.

Now, back to the main story of the Alcon blue butterfly. The butterfly is found in alpine meadows of Switzerland and the Caucasus Mountains to the east. Some larvae of the butterfly use a remarkable tactic after their fourth molt: they drop to the ground and wait to be discovered by common ants in the genus *Myrmica*. Amazingly, these ants readily carry the caterpillars to their nests, where the impostors are nursed and cared for until they pupate and metamorphose into butterflies.

Butterfly caterpillars have a much larger appetite than ant larvae. Incredibly, nurse ants often neglect their own broods to provide food for the alien species.[33] How could this happen? The secret is olfactory fakery—caterpillars can imitate the smell of ant larvae. They also mimic the noises and vibrations typical of the queen ant. This allows them to raise their status and thus their priority for receiving food and care.[34]

These tricks allow the caterpillars to literally live like royalty in a colony of ants (fig. 3.6).

How does the Alcon blue butterfly larva pull off this trick? Ants produce a blend of chemicals called hydrocarbons and rub it like a lotion onto their body. It serves as an olfactory badge unique to that particular colony. Ant guards and soldiers admit anyone who carries that "colony badge" into their nest, which is otherwise defended like a military fortress. Meanwhile, anybody who bears a different badge would be kept out or even killed. The ants' recognition system is so specific and automatic that even if you paint a glass or plastic dummy with chemicals matching their colony odor, they'll happily carry it into their nest.[35] This is exactly how the butterfly con artists obtain security clearance for ant nests: they fake their identity by producing the right chemicals. These butterflies have deciphered the secret code of their ant hosts.

Masquerading as ants, however, isn't free of risk. In fact, many caterpillar impostors are killed soon after they're discovered and carried into the ants' nest. This selective pressure causes the caterpillars to evolve chemical blends that can better fool the ants so they can get through the security check. Of course, the ants can be expected to fight back, especially when impostors are common. Think about what you would do if swindlers and scammers were everywhere.

Indeed, colony odors in some ants have evolved to be progressively specific until they are extremely difficult to be faked. This is paralleled in the digital world by our need to use complex passwords to gain access to our accounts. Whenever there is a need for security, we wouldn't use "password" or "1234." We instead choose longer and more complex ones peppered with rare characters and symbols, for they are harder to guess. This is exactly what the ants do as the pressure from cheating butterfly larva grows. As the screening process for chemical passwords tightens, many less skillful mimics are rejected and killed. Then, the butterfly caterpillars answer in kind by evolving more accurate fakes. The evolutionary arms race between cheating and cheat-detection never ends, and the host-parasite relationship becomes increasingly specialized for each other.

Not all Alcon blue butterflies live as con artists, though. But for those that do, they face another problem. How can they be discovered by the

FIGURE 3.6. (a) An adult Alcon blue butterfly (photo credit: Rob Zweers) and (b) its parasitic caterpillar carried and cared for by an ant (photo credit: David Richard Nash).

right ants among many in the same neighborhood? From the perspective of a butterfly caterpillar, the probability of being picked up by a particular species of ant is quite low. So, you shouldn't put all your eggs in one basket. You have to cast a wider net to increase your odds of success.

For this reason, some caterpillars produce hydrocarbons that mimic the identities of several different ant species so that they're more likely to be picked up. But this cast-a-wide-net approach can cause another problem. The ants on which your entire life depends can be fussy. If you smell slightly different from the odor they prefer, they may lose their enthusiasm or even reject you. Facing this problem, caterpillar impostors have come up with an ingenious solution: alter their odor to match the host colonies *after* they are picked up. They do this either by producing the chemicals themselves or by rubbing their bodies with the colony scents of the ants.[36] The ability to manipulate their body odor gives the caterpillars the flexibility to gain entry to ant colonies and the specificity needed to succeed once they're inside.

Although often defrauded by other animals, many ants themselves are not innocent of cheating scams. The best examples are ants known as slave-makers. These ants use a combination of brute force and chemical mimicry to capture ants (usually of a different species) to work for them. Some slave-makers are so dependent on their slaves for food and care that they have lost the ability to live by themselves.

Although situations differ somewhat across species, workers of slave-making ants are mostly specialized pirates. Their job is to raid other ant colonies, slaughter their workers, and steal their victims' grubs. In some cases, marauding slave-makers use chemical warfare, spraying special compounds stored in a tiny bag, called the Dufour's gland, inside their bodies. They're like soldiers carrying tanks of poison gas. Some of these chemicals have potent psychotic effects. They can disarm the fighting morale of their target ants and induce a panic that makes their rivals flee en masse. Even worse, these gas attacks can cause their victims to turn on their own colony mates and fight each other to death.[37] Armed with these chemical weapons, slave-makers can steal thousands of broods from other ant colonies in a single season. They then brainwash their slaves by mimicking their body odor or by infusing the slaves with the masters' odor instead.[38]

꙰

When we talk about mimicry, especially Batesian mimicry, we often refer to animals' ability to make use of static objects such as plants. But the tables can be turned. The Second Law of Cheating still applies, but in this case, the brainless species succeeds in duping those with brains.

As we brought up in chapter 1, nearly a third of all orchids are pollinated by practicing deception of some sort, that is, by fooling insect pollinators into trying to mate with flowers posing as female insects. Considering that orchids are the most diverse family of flowering plants, with a total of no less than 20,000 species (about 7% of all known flowering plant species), the prevalence of impostors in orchids is staggering. In other words, they cheat to take advantage of a free fertility service on a grand scale, using the Second Law of Cheating.

Achieving fertilization by tricking pollinators into copulation is hardly the only type of deception employed by plants. Many use deception to hide from or deter their nemeses: herbivores. Stone plants in the arid areas of southern Africa mimic pebbles so as to conceal themselves. The tasty leaves of the yellow archangel and dead nettle imitate those of the stinging nettle, whose tiny thorns can cause skin lesions. These tricks can cause herbivores to leave the impostors alone.

Passion flowers in South America may win the prize for the most peculiar form of plant mimicry. These plants can "read the mind" of *Heliconius* butterflies, whose caterpillars can cause extensive damage to the leaves. However, the butterflies avoid laying eggs on leaves that already have butterfly eggs for a good reason: to give their offspring enough to eat. Because of this behavior on the part of the butterflies, passion flowers have evolved growths on their leaves that resemble butterfly eggs (fig. 3.7). In this way, they discourage butterflies from laying eggs and damaging their leaves.[39]

If you are a mushroom enthusiast like me, you may have noticed that different species of mushrooms often have distinct scents. And different odors attract different species of insects, such as fungus gnats,

FIGURE 3.7. Egg mimicry in passion flowers: the yellow spots on the leaves resemble butterfly eggs, which deters butterflies from laying eggs that would hatch into caterpillars that would feed off the leaves (advertisement in Grassy Knoll Exotic Plants online catalog; used with permission).

that mushrooms rely on to spread their spores. But odors are not the only lure mushrooms deploy to attract insects. Some mushrooms, such as the jack-o'-lantern, glow in the dark to attract insects that serve their needs. And, as we learned in chapter 1, truffles can entice pigs to spread their spores by producing fake pig pheromones.

Carnivorous plants, such as pitcher plants, sundews, and Venus fly-traps, lure insects by using nectar or by imitating the color and/or odor of flowers or other insect food. The giant *Nepenthes rajah* pitcher plant in Malaysian Borneo produces a pitcher that can grow to 40 cm high and 20 cm wide. It can contain over a half gallon (1.9 liters) of digestive fluid, enough to trap and break down a mouse. Some tropical pitcher plants glow under ultraviolet light to attract insects.[40] They can even resist the "temptation" to eat insects immediately. Instead, they can temporarily deactivate their trapping device so as to lure more ants into their trap before closing it.[41] Such "delayed gratification" is not easy, even for humans.

Some plants can fake their own death. For example, sensitive plants (*Mimosa*) play possum when you touch them. In nature, their worst enemies are herbivores such as grasshoppers, which love to eat fresh leaves. However, touching a juicy leaf of the plant triggers a series of reactions in specialized cells. This generates a weak electric signal, like neural signals in animals. The signal is relayed to the leaf and leaf blades, which then pretend to wilt, causing the befuddled grasshopper to lose its appetite.

✞

In the previous chapter, we learned that a common method of cheating, which we dubbed the First Law, works by falsifying messages in signals. It works well for cheaters during communication between members of the same species. Between different species, however, communication is far less common. So, in most cases, the First Law doesn't apply— except when one species can break the communication code used by another species, as we've seen with the eavesdropping *Photuris* firefly, the bluestriped fangblenny, and the Alcon blue butterfly.

In this chapter, we've ventured into the world of cheating between species and discovered a new method we've named the Second Law. It refers to deceptions that exploit cognitive loopholes in the target of the cheat. We've seen that animals, plants, and fungi can all use the Second Law to victimize other animals, many of which have cognitive systems far more advanced than their victimizers—which, in the case of plants and mushrooms, have no cognitive systems at all. Through a series of examples, we've gotten a glimpse of how an evolutionary arms race between the cheating and the cheated can lead to a wide range of sophisticated phenomena such as mimicry, camouflage, bluffing, faking death, and other tactics used by plants, fungi, and animals.

As we have seen repeatedly, honesty is often exploited by members of the same species and by members of different species. Why, you may be wondering at this point, does honesty even exist? The answer is that honesty, like cheating, provides adaptive values that can also increase Darwinian fitness. Although this might seem counterintuitive, honesty is a particularly successful strategy when there are lots of cheaters around. To see how honesty evolves and flourishes, we turn to the next chapter.

CHAPTER 4

Infidelity and the Rise of Honesty

One summer day in 2009, Shine, my then ten-year-old son, was playing golf with his friends on an empty lawn outside our house. Suddenly, their playful laughter stopped, and an eerie silence fell, as a stray golf ball flew over the fence and smashed the back window of an older house with peeled red paint, visible from our backyard. All the kids immediately scattered—except my son. One boy turned and yelled, "Run, Shine! Run!"

Shine didn't move. He stood there for a moment and then walked over to take a close look at the broken window. Inside the house, a fuming older couple were scolding and cursing. Shine went to the front door, knocked, and when the owners came to the door, explained what had happened.

Just when I was about to walk out to check on the situation, the woman called me, explaining the situation. There was no trace of stress or anger in her voice, so I was relieved. "We'll let you know how much it costs." She hung up.

The bill came a few days later: $198. The couple delivered the bill personally, praising Shine's honesty while insisting that they pay half of it. "Such a good kid!" They left, smiling. They were kind: the bill was more of a burden to them than to us.[1]

The golf incident has since become a legend for the Sun family. For me, it was also an instructive illustration about the price of honesty.

How could honesty survive and thrive under the pressure of its cost in the biological world? One simple answer is to *make cheating pay*. If cheating is costly enough to the cheater, then the dictum that "honesty is the best policy" becomes closer to truth.

Nature, indeed, has evolved many ways to accomplish this. We have already seen some examples in previous chapters. Many species of ants, bees, and wasps, for instance, use a policing system to detect and destroy eggs surreptitiously laid by workers. As expected, when enforcement is lax, some workers can get away with laying illicit eggs.[2]

Cheating by free riding, however, is often harder to detect and quite difficult to punish. For example, it's difficult to know whether the wolf farthest back in the hunting pack is there by chance or is dodging its duty. That's why free riding is such a ubiquitous problem in animal and human societies. However, if the impact of free riding becomes so substantial that it threatens a communal undertaking, it can trigger the rise of policing and punishment. This is exactly what biologist Lee Alan Dugatkin found in small fishes like guppies and sticklebacks.

Guppies and sticklebacks are snacks for many predatory fishes. How can guppies and sticklebacks hunt for their own food without becoming food themselves? Evolution came up with a solution, and it feels like a strategy written in Sun Tzu's *Art of War*: "If you know yourself and know your enemy, you will never lose a battle in a hundred." That is, to defeat your enemy, you need to spy on him. But for a little fish, approaching a formidable predator to collect information is like flirting with death. The chance that a fish on such an espionage mission will survive for 36 hours is less than 50%, compared with that for a fish who stays put.[3] How can a fish lower the risk?

The answer turns out to be by banding together, taking advantage of safety in numbers. Even two fish spying together can cut the per capita risk in half. This strategy, however, has a minor flaw: the fish who leads the way exposes itself to a higher risk. The lead fish's risk is even greater if others fail to follow through on the cooperative effort, leaving the leader to its own devices. Clearly, natural selection doesn't favor the leader but rather those who say, "You go first." But if none of the fish were willing to take the lead, the spying mission would be dead in the water, so to speak. To make the joint venture viable, the risk must be evenly shared among the fish in the gang. How can they handle this dilemma?

Guppies and sticklebacks, despite their tiny brains, have come up with an ingenious solution. When two fish go to inspect a predator, the leading fish may make a sudden turn, switching places with the trailing fish.[4] As a result, the trailing fish becomes the leader. By swapping roles, the risk associated with leadership is shared. This behavior can be seen as a form of enforcing compliance with the cooperative effort, ensuring that the trailing fish can't shirk its responsibility to share the risk.[5]

Guppies and sticklebacks only show a small set of many methods used to impose costs on cheaters. In the remainder of this chapter, we'll look at a wide range of strategies that animals use to control and confront cheating. From these examples, we'll seek a deeper understanding of the ways that evolution promotes honesty in the natural world.

𝕏

Let's start the tour in one area where cheating is rampant and also resonates with our human experience: reproductive relationships between males and females. Sex cheats are widespread in animals and can range from mildly intriguing to utterly bizarre. And, as a result, infidelity is often a major problem. Why?

The short answer is that males and females, despite an equal genetic share in their offspring, have different reproductive interests at stake. What is beneficial for the male may not necessarily be beneficial for the female, and vice versa. To understand the root cause of this conflict of interest between the two sexes, we have to travel back to the time when the first male and female appeared out of the blue about two billion years ago.

At the time, pretty much all forms of life were busy reproducing asexually, that is, genetically cloning themselves. This allowed them to multiply fast and spread far. But despite cloning's high efficiency, it has one major shortcoming—it can't add genetic diversity. An entire lineage could easily go extinct due to the combination of bad mutations building up and relentless attacks by parasites and pathogens. This might have been the inevitable fate of eukaryotic organisms—until two single-celled protists merged themselves first before producing daughter cells.[6]

This seemingly trivial occasion was a revolutionary event in the history of life: it marked the origin of sex. Like shuffling a deck of cards before you start a new game, sex shuffles the genetic makeup every time a male and female mate and produce offspring. In doing so, sex kills two birds with one stone—the accumulation of bad mutations and vulnerability to parasites and pathogens. Unfortunately, this epic event had unforeseen consequences. It drew males and females into two perennial battles over sex: in one battle males compete against females, and in the other battle males and females compete separately against others of their own sex.

The first long battle was fought over a seemingly petty issue: the size of reproductive cells, or gametes. The cause was rooted in a free riding cheat. At the dawn of sexual reproduction, the two gametes that fused were equal in every respect: size, genetic stake, and material and energy invested. But before long, one gamete discovered that it could gain an evolutionary advantage by cutting corners—by contributing a bit less material and energy into each gamete. This could allow it to make *more* gametes, which in turn meant more opportunities to merge with the opposite gametes. As a result, its genetic return—its own fitness—was increased. This fitness advantage forced others to follow suit—to jump on the free riding bandwagon themselves. Consequently, more and more gametes were churned out over time, each becoming ever smaller with less and less material and energy stored in it. Finally, little was left except the genetic package plus a few mitochondria, devices for the power generation that is essential to moving the gamete for the purpose of fertilization. At the end of this process, the impoverished little gamete is now labeled a sperm, and its maker, a male.

The gold rush for small, cheap, and numerous sperm had its downside, however. As the investment in each gamete was reduced, the mortality of its desired product, the zygote, went up. This created a new opportunity for another type of gamete to thrive by going in the opposite direction: raising their chance of survival by storing more material and energy reserves. As a result, these gametes became eggs and their creators, females. (If a chicken egg is still unimpressive in size, try an emu or ostrich egg. One is more than enough for a big meal for both you and me.[7]) So, when this long battle of the sexes drew to a close, there was no clear winner. Instead, a compromise was reached—an evolutionary Magna Carta of sorts—declaring male and female as equally successful strategists in sexual reproduction.

Peace? Not yet! The second epic battle of the sexes broke out after males and females emerged and is one where males and females compete separately. This battle was fought over who could contribute more copies of genes *within their own sex*. Although sperm are small and numerous and eggs large and few, exactly one sperm and one egg are needed to seed a new life. For this reason, there is no net genetic gain when you pit your fitness against the fitness of your peers of the *opposite* sex. An alpha male macaque may snatch a banana from a female by force, but he gains no benefit to his fitness by antagonizing her. This is because his fitness advantage is relative, in comparison with other males. His edge comes mainly at the expense of other males, not females. Just as a male sprinter can't win the 100-meter race by running faster than the female champion, neither can a male (or female) win the evolutionary race by increasing their fitness in relation to the opposite sex. This second battle of the sexes is like most sports, in which men and women compete separately.

The enormous gap in reproductive potential led to a marked contrast in the strategies employed by males and females in their quest to maximize their individual fitness. Ismail Ibn Sharif, the sultan of Morocco, made it into the *Guinness World Records* by siring 877 children. He accomplished this feat by ruthlessly maintaining a large harem, earning him the nickname "The Bloodthirsty." The accomplishments of women who made it into *Guinness* pale in comparison to Ismail, although their feats are still amazing. Two eighteenth-century peasant wives of Feodor Vassilyev, in a little-known village in Russia, gave birth to a total of 87 children. The first woman gave birth to 16 sets of twins, 7 sets of triplets, and 4 sets of quadruplets for a total of 69 children, and the second woman gave birth to 6 sets of twins and 2 sets of triplets for a total of 18 children. Out of the 87 children, only three died in infancy, well below the average mortality at the time.[8] Although little information is available, the two women earned their place in history by having enough to feed the children plus an unusual genetic boost apparently from their husband.

These anecdotal world records in human fertility reveal a general trend in the animal world: male fitness tends to rely on how many females they manage to mate with. Consequently, their game plan is usually based on gaining access to as many females as possible, whatever it takes. For females, by contrast, mating with large numbers of males is unlikely to do them any good, as long as they have plenty of resources to rely on for raising their children. Since females make a small number of expensive eggs, their winning reproductive strategy revolves around finding males like farmer Vassilyev, who could contribute sufficient resources and/or good genes for their offspring. Therefore, a rule of thumb for adaptive sexual behavior is that females follow resources and males follow females. This is known as Bateman's rule, named after Angus John Bateman, who discovered the pattern in fruit flies in the 1940s.

This disparity in reproductive strategy shows why females can be penalized for being too rash in choosing their mates. Males, on the other hand, can suffer if they don't take advantage of every opportunity to mate. That's where the classic sexual stereotypes come from: fussy females who wait to find just the right male, and males hard at work courting every female they can find by any means necessary including charming, cajoling, and coaxing.[9] To gain an advantage during the process, both sexes are ready and willing to cheat. That's why lying and deceiving are widespread in sexual reproduction. Because there is always the possibility that cheating will elicit counter-cheating, sex cheats hold the secret for the evolution of honesty. That's why they deserve our special attention here.

❧

When there is infidelity in a monogamous animal, males are usually the first sex that comes to mind. But despite male notoriety for being unfaithful to his spouse, females of many species are known to betray monogamous relationships as well. And there is no reason why females should remain blindly loyal to their males when the interests of the two sexes are different. Indeed, females in many species will practice deception for sex when it promotes their own fitness. And they often do so when they are most fertile. This is commonly observed in birds.[10]

Traditionally, birds—especially songbirds—were thought to be largely monogamous because most of them form reproductive pairs. In the 1980s, this long-held belief began to falter. A powerful new tool—DNA fingerprinting—enabled researchers to determine which males the females have mated with. We now know that around 90% of pair-bonding birds frequently engage in extra-pair copulations, mostly by cheating behind their mates' backs.[11] Clearly, birds are only socially monogamous, not genetically.

Social monogamy is far rarer in mammals than birds. Extra-pair copulation is found in many pair-bonding mammals. In marmots, for example, a colony is typically made of an adult pair and their offspring. Genetic paternity testing shows that 20% of the offspring are actually sired by males from other colonies.[12] In many species of monkeys, likewise, high-ranking males are supposed to monopolize mating opportunities with females. Nonetheless, females often engage in sex with low-ranking males behind the dominant male's back.

Why is extra-pair copulation common in birds and mammals? The apparent answer is that eggs in these animals are internally fertilized, inside the female's body. Unlike external fertilization in most fishes and amphibians, where males can see their sperm meeting the eggs, internal fertilization creates a time lag between mating and the actual union of sperm and egg. This enables females to mate with multiple males and choose sperm from preferred males to fertilize their eggs. Males, meanwhile, can be kept in the dark about a female's sexual history, which adds uncertainty to their fatherhood. This information asymmetry in offspring paternity gives females a significant advantage in their capacity to deceive males with extra-pair copulation. But why do females cheat? What is the benefit they gain from these affairs?

Some biologists think that extra-pair copulation in females is the evolutionary side effect of males seeking cheating opportunities.[13] However, most believe that females have something to gain from these clandestine relationships. Sexual cheating may allow the females to have access to additional resources, such as food or shelter, in another male's territory.

In many situations, however, females get no direct reward from extra-pair copulation. What then drives females to sneak around? There are three main reasons. One is that females may form their primary pair-bond with a male who presumably possesses genes of inferior quality. By pursuing a covert liaison with a more attractive male, females can endow their offspring with a better genetic inheritance.[14] This might enable their children (both sons and daughters) to survive better or make their sons (but not daughters) more sexually appealing to females. In the socially monogamous dark-eyed junco, for example, birds born out of "wedlock" can rack up 85% more offspring than those produced by the monogamous pair. This case illustrates that females can indeed be richly rewarded for surreptitious sex by producing more grandchildren.[15]

A second advantage gained by females who engage in illicit sex is fertility assurance. As we know, eggs are few and are expensive. If they are not fertilized in a timely manner, females can lose big, even missing an entire reproductive cycle in many species. In seasonally breeding birds and mammals, they could miss an entire season, which typically means a whole year. Such a loss is not easy to make up. That's why in some pair-bonding species, such as the swan, a couple may split up and find new mates if their initial reproductive efforts fail. This is more frequently seen among short-lived species since they face a tighter deadline. By mating with several males, a female is more likely to get her eggs fertilized—in case the first male she pairs with is infertile or genetically incompatible.

The third reason for a female to cheat is to have genetically diverse offspring sired by different fathers. This allows her to avoid the risk of putting all her eggs in one basket. It's no coincidence that diversification is also a basic principle in financial investment. You can lose everything by betting all your money on a single company that fails. Likewise, animals can lose everything by remaining genetically homogeneous, which can reduce the immune capacity needed to fight a broad range of germs and parasites. The risks entailed in monoculture crops have taught us much about the importance of genetic diversity. One well-known example is the Irish famine that resulted from potato blight. More recently, banana plantations have suffered high losses due to Panama disease, caused by a fungal infestation that can devastate crops for lacking genetic diversity. So, diversification is as important in evolution as it is in agriculture and finance. It's a potent tool to hedge the risk of losing everything.

Female infidelity, however, has its downside as well. Cheating females, if discovered, risk retaliation from their cuckolded mates. In many species of birds (such as the barn swallow mentioned in chapter 2), when a male sees his "wife" attempt to mate with another male, he punishes the female by violently attacking her. In other species, males may withhold their parental contribution or simply abandon the nest, leaving the females to hatch and raise their chicks on their own.[16] If that happens, cheating females can suffer egregiously, due to higher chick mortality or even the demise of the entire clutch.[17]

In mammals, males have a wider variety of means to prevent females from cheating. The severest of all is infanticide, killing offspring that are "illegitimate"—in other words, not theirs. Male infanticide in primates is extremely common—so common that for every ten infants born, more than three in gorillas and six in langurs fall victim to infanticide.[18] Since infanticide is devastating for females, pregnant females may take a variety of steps to prevent it, to a degree.

One such strategy is to terminate their pregnancies rather than carry them to term. In small rodents, for example, a female in early pregnancy can abort the embryos in her uterus when she sniffs the odor of a male other than her mate, a phenomenon known as the Bruce effect, credited to biologist Hilda Bruce, who discovered it in 1959.[19] In the chemical language of rodents, the presence of a strange male's odor is interpreted by a female as the defeat or death of her mate. This portends an inevitable infanticide by her new mate down the line. Abortion, though harmful enough for her legacy, is still better than losing already-born pups. By ending a losing investment early, females can avoid the gambler's fallacy (also known as the sunk-cost fallacy[20]) and salvage precious time and energy for a more promising pregnancy to come.

How can males become aware of female infidelity? One obvious clue is the timing of birth. But do they have other means as well? The answer is yes, and I was lucky enough to have the opportunity to discover one such method myself for my PhD work, under the guidance of my advisor Dietland Müller-Schwarze.

In the early 1990s, I live-trapped beavers at Allegany State Park in western New York and collected their pheromones from the castoreum pockets and anal glands of more than 300 beavers.[21] I then carefully tallied the types and amounts of chemicals from these secretion samples using gas chromatography and mass spectrometry. This information allowed me to compose a chemical profile, like a facial image, to represent the identity of every beaver I'd trapped. I then compared these chemical images, as we do with family photos, to see whether close relatives share more similarities than more distant ones. (DNA work was not done at the time. The relationships among family members were constructed based on more than a decade of continuous trapping and observation data on individually identifiable beavers.) The moment that the first positive sign emerged, I was completely beside myself. I was so elated that I worked nonstop through Christmas Eve that year in the Syracuse lab of my mentor and friend Steve Teale. (Warning: science can be highly addictive!)

Sobriety returned as the ecstasy subsided. While the genetic information about kinship was clear, it didn't tell whether beavers were able to use it to their advantage—a question that was yet to be explored. So, I decided to test this in the field. It took me three years to gather enough data to come to a verdict: beavers could indeed use anal gland secretions to sniff out relatives, as opposed to strangers, and to discriminate between close relatives and distant relatives as well. I later dubbed these chemicals kinship pheromones (fig. 4.1),[22] which had been known to be used in kin recognition in other animals such as tadpoles.

It turns out that beavers aren't the only mammals that use kinship pheromones. Giant pandas and some small rodents also produce

FIGURE 4.1. Beavers typically form monogamous pairs and assess genetic relatedness by
kinship pheromones (photo credit: Dietland
Müller-Schwarze).

kinship pheromones, according to studies done in the labs of my collaborators Jianxu Zhang and Dingzhen Liu.[23] These findings are likely
a scientific harbinger that hints at a bigger picture to be unveiled: kinship pheromones may exist broadly in animals, especially those with a
sharp olfactory sense. Unfortunately, simian primates—the group to
which we belong—have a relatively poor sensitivity to odors.

Why is it critical for animals to know their genetic identities? From
a traditional viewpoint, it enables animals to solve two major challenges.
One is the practice of nepotism, common in the animal world. Knowing
genetic identity helps them help their relatives, rather than wasting
costly altruism with random acts of kindness toward total strangers. The
other is for mate choice because a pair that is genetically too closely or
too distantly related may lower the fitness for the pair.[24] Kinship pheromones can certainly fill these two needs.

But knowing the genetic identity of their offspring has another vital function regarding female infidelity. It allows males to protect their paternity by preventing female partners from engaging in extra-pair copulation. This is particularly important in species like beavers, where males and females form long-term pair bonds and males invest heavily in caring for the young. A male may lose a great deal, in terms of his investment in promoting his own fitness, if his offspring are the results of his mate's covert liaisons.

That was illustrated by a study done in Russia, in which paternity testing using DNA fingerprinting technology showed no sign of extra-pair copulation in the Eurasian beaver.[25] An apparent contradiction came from a study of the North American beaver in southern Illinois. Using a similar paternity testing method, Zhiwei Liu and his research team found that some young beavers from the same colony were sired by different adult males.[26] Was this the evidence for female extra-pair copulation? The answer is most likely no. Beavers may no longer stick to a typical nuclear family made of an adult pair and their children when living in conditions of high density. Instead, they often form complex families with three or more breeding adults as a response to varying ecological conditions. Moreover, families may also allow close relatives from other colonies to stay for a while, like guests. It's unsurprising that young beavers born in the same colony have different fathers.[27]

🐾

There is an old Chinese saying: "A man should be worried about having the wrong career; a woman should be worried about marrying the wrong man." Although this idea is obsolete for modern human life, few decisions are still more consequential than choosing the right mate for many female animals. Though capable of saying sweet things during courtship, males vary a great deal in quality, nonetheless. Some are good fathers; others not so much. Some have good genes that can make the next generation stronger and more attractive; others are poor in both regards. Although females in species where fertilization is internal have

FIGURE 4.2. A music frog male with "property"—a hole
(photo credit: Yue Yang).

an edge in the information warfare regarding paternity, they don't have
the upper hand when it comes to the quality of males. How can females
tell the real deal from an apparently attractive phony? One way is to rely
on male traits that are hard to fake.

In the music frog, for instance, males arduously dig holes in the mud
to make nests—their investment in real estate. They broadcast their love
songs from inside these holes to attract female attention. Since the male
nests also provide protective shelter for eggs, 70% of female frogs fall for
males who have a nest over males who don't. Females are able to distinguish males with a nest from those without by the frequency change in
male songs emitted inside versus outside a hole (fig. 4.2). A homeless male is unable to fake that desirable quality—he can't croon a love
song to claim real estate he doesn't in fact own.[28] Female parakeets are
also attentive observers of male behavior. They pick males that have
demonstrated puzzle-solving skills, which may indicate intelligence
that is essential for surviving complex or unpredictable environmental
conditions.[29] Both examples demonstrate how females choose males
based on behavioral traits that directly or indirectly benefit their offspring. And more to the point, these traits are hard to fake.

Historical information about Feodor Vassilyev's life is scanty. But if his track record of reproductive success can tell us anything, it's that he had to have adequate resources (and maybe a good genetic makeup as well) to have set a world record that is difficult to match. This is as obvious to us now as it was to the villagers of his time. Generally speaking, information like ownership of resources and territory in humans, or nest possession in the music frog, and problem-solving ability in the parakeet, is known or advertised in some way. This provides both males and females with same knowledge, making the mating game symmetrical from an informational point-of-view.

More often than not, however, essential information about male quality—the potential for future success, likelihood of paternal care, the superiority of genes, and so on—is at best only partially accessible to females. For this reason, it can be readily faked. Females are thus at a disadvantage, forced to play an asymmetrical mating game with incomplete information, giving males a distinct edge.

Keep in mind that the evolutionary pressure for males largely lies in mating with as many females as possible, as Bateman's rule posits. Resorting to deception is a means to an end: gain access to females as quickly and cheaply as possible. We are quite familiar with the same type of situation in our own species. For example, a penniless college student may borrow a BMW convertible to attract a date. In the case of *The Great Gatsby*, Jay Gatsby concocts a major bond fraud to impress the naïve Daisy with an illusion of vast wealth. Although F. Scott Fitzgerald's story is fiction, the ruse is real—all too real, in fact. We've all heard of similar scams.

Choosing quality males is even more important in animals where males contribute little more than their genes to the reproductive process. In animals such as peacocks, moose, and many species of rodents, males offer neither nuptial gifts nor parental care. Females can do nothing but try to pick males who have good genes during the courtship process. Now, assume you're a female of such a species. Your task is to spot the fittest dude from among a large bunch who are all trying to convince you: "I am the one!" How can you tell the real deal from a phony?

Let's consider the worst scenario first. If you do fall for the proverbial "wrong man," you will suffer a major setback in your fitness—one from which you may not recover. This puts you at a clear disadvantage with respect to the males. This scenario suggests what evolution might do for females: reward them if they can come up with counterstrategies that enable them to distinguish a phony from a truly high-quality male. Female "coyness" is just such a tactic. It evolved to prevent them from jumping to conclusions too soon without adequate vetting, making females far less impulsive than males, especially when it comes to mating. Therefore, the stereotype of "coy female" exists for a reason. It gives them time and leverage to locate Mr. Right. (We should be keenly aware that stereotypes are extreme simplifications of complex realities for the convenience of reasoning. In fact, males and females in few animal species would entirely comply with Bateman's rule.)

Still, coyness can't inform females which male is lying and which is telling the truth. Females need a more effective tool—a powerful lie detector—if they hope to find out who's a high-quality male, and who's a fake. How can they do this?

Enter the tale of the peacock's tail. Most of us are awed by the peacock's stunning plumage, particularly the long tail studded with iridescent eyespots. Darwin, however, had a much different reaction to this phenomenon. In 1860, one year after the publication of *On the Origin of Species*, he was still grappling with the question of why evolution might lead to such an exaggerated structure. In a letter to his American friend, botanist Asa Gray, he confessed sheepishly, "The sight of a feather in a peacock's tail, whenever I gaze at it, makes me sick!" Why? Because he couldn't figure out how the peacock could have such a showy tail that's completely useless for—and actually detrimental to—its own survival. We can understand how frustrated he was when his grand theory of evolution by natural selection apparently didn't apply in this case.

The peacock dilemma tormented Darwin's mind for twelve more years until the publication of his second masterpiece, *The Descent of Man*. In this book, he settled on the idea that animals have an innate sense for what is attractive, or as he put it, "a taste for the beautiful." Even so, Darwin wasn't entirely satisfied, because he knew little about how the process works. Why do peacocks bear the large, showy, and seemingly useless tail that made Darwin feel sick?

Given that the peacock's tail is used for nothing but attracting females, the question becomes: Why would females fall for something as useless as that? This dilemma had become one of the greatest puzzles in science. And it went unsolved for more than a century until 1975,[30] when the Israeli evolutionary biologist Amotz Zahavi came up with an intuitive idea called the handicap hypothesis, also variously known as the indicator, good genes, costly, or honest signaling hypothesis or principle.[31]

Zahavi's logic is simple: a peacock who bears such a tail has to be biologically superior. Otherwise, he would have already succumbed to the scarcity of resources and abundance of predators. So, a large tail is itself a manifestation of male fitness, an animal version of conspicuous consumption. Wimps and phonies are unable to afford indulging in such a luxury. Thus, the handicap hypothesis also neatly explains such puzzles as why male guppies sport orange spots on their bodies, why male birds in many species display beautiful plumage, and why male elk grow huge antlers. Because these vanity ornaments require a lot of material and energy to produce yet impose a high level of risk to their bearers, they are a litmus test for male quality: those who can afford them are genetically fit; those who can't must be inferior.

With such ornamental handicaps serving as hard evidence, males are unable to lie about their quality, even though they want to. Information warfare is thus no longer asymmetrical, and females are empowered to play the mating game with males on level ground. This shows how handicaps, even though many appear useless and wasteful, serve the useful purpose—at least from a female's perspective—of exposing male honesty.

The handicap principle makes perfect sense in the framework of our own experience. If you are mentally gifted, for example, you still have to demonstrate it by showing superior abilities in handling difficult tasks such as playing a musical instrument with virtuosity, reciting Shakespeare's verse flawlessly, solving differential equations, or better yet, a combination of all. Such skills may be mostly useless in daily life, yet they take time, determination, and most of all, a high level of intelligence to perfect. These skills are widely admired, mostly not for their practicality but for their honest demonstration of cognitive capacity, which may bode well for future success. (That's also part of the reason many companies want job applicants to provide evidence of college degrees, even SAT scores or GPAs.) For the same reason, animals bear ornamental plumage, sing songs, perform dances, or show their ability to solve puzzles. These traits and skills are burdens (at least in part) that enable females to distinguish high-quality from low-quality males. A handicap trait, in a nutshell, acts as a lie detector.

๙

Handicaps come in many varieties. Some are obvious, such as exaggerated ornaments in birds like the long tails in peacocks, pheasants, and widowbirds, not to mention birds-of-paradise (color plate 10). Some are subtle or hidden—a high testosterone level, for example. It is a handicap because it compromises the immune system, acting as a poison in the body. The message is clear: if you can afford that much poison in your bloodstream, you must be fit.

Most handicaps are part of the body. Bright plumage in males, for instance, is a good indicator of an immune system protective against blood-sucking parasites.[32] That's why females of many bird species prefer males with gaudy feathers. However, handicaps can also be external in some instances—that is, they can be outsourced to something else. For example, bowerbirds spend a vast amount of time and energy meticulously building and perfecting decorative structures called bowers that serve no purpose (it is not used as a nest) except to attract females (see fig. 5.8 in chap. 5). That is, the bower serves as a handicap that indicates male quality.

Some handicaps are not only subtle, but unperceivable to humans as well. I learned this in the 2000s, when my longtime collaborator Janxu Zhang at the Institute of Zoology in Beijing tested whether mice could sense stressed versus nonstressed peers. If females could, we expected that they would prefer unstressed males because a stressed mate would indicate a less-fit mate. We tried a simple experiment: get two groups of males, let one sniff cat pee (a stress-inducing odor for mice) and the other rabbit pee (not stress-inducing) for eight weeks. We then allowed females to choose between the two groups of males. To our great surprise, females could not only tell apart the two kinds of males but also prefer males who had experienced stress by being exposed to cat pee. This was the opposite of our expectation.

We scratched our heads for quite some time, mulling over what was going on. Then one day, during a chat over lunch, we were struck by a simple idea. To understand the peculiar observation, we need to put ourselves in mice's shoes. Their world is a struggle for life on a daily basis because vast numbers of mice fall to predators such as wild cats, weasels, and owls. So, if you're a female mouse, it's critical to find the fittest male among many, all screaming "I am the one!" when they flirt. What can you do? You surely can't give them the rodent equivalent of an IQ test. But if you can determine a sign that honestly indicates the fitness of a potential mate, that would help you spot Mr. Right.

It turns out the sign you are looking for is in the pee of the male mice. The logic? A number of hormones in mice who have just experienced a stressful event are noticeably different from those who have not. These changes will show up in their urine as metabolic waste products, which coincidentally serve as a truthful record, like a GoPro video clip, telling the story of their thrilling encounter with danger, most likely a predator. Females can pick up the message when sniffing the urine of such a male: he has just heroically returned from the mouse battlefield, where many of his comrades likely fell prey. Because of the signature of stress hormones in his pee, his very existence speaks eloquently about his quality as a mate. That's why female mice swoon for males that have had prior experience with predators. To summarize the essence of our finding, journalist Charles Choi, in his article in *Live Science*, simply stated that cat urine "makes male mice more macho."

You may wonder what's in the urine to make the male macho. We looked into this question as well and nailed down the signature changes. The levels of four male pheromones—E,E-α-farnesene, E-β-farnesene, R,R-dehydro-exo-brevicomin, and S-2-*sec*-butyl-dihydrothiazole—are higher after a stressful experience.[33] Don't worry if you're unfamiliar with the names of these esoteric compounds. But the logic should be clear: these chemicals work exactly like the peacock's tail—the more stress hormones, the fitter the mouse.

For a trait to serve as a handicap, it must be costly to produce and maintain because being costly is a means to an end for honesty. However, this would place males under a constant and heavy burden, making them vulnerable to predators and pathogens and lowering their chance of survival. Can males cut corners, using handicaps only when needed most? This would give them the best of all worlds.

The answer is yes. In many cases males can reduce that burden by taking on handicaps only when they're needed. For example, deer antlers are costly to bear in terms of material, energy, and risk involved, but are useful in the mating season, when they serve the dual purpose of attracting females and defending their turfs or harems. When the mating season is over, the antlers are shed to free males from being hampered by the heavy load. Similar cost-cutting measures are also seen in white pelicans, which grow a "horn" on the top of their beaks (in the American white pelican, fig. 4.3a) or a "bump" near the eyes (in the great white pelican, fig. 4.3b). These structures demonstrate that the pelicans can still do well when their vision is partially impaired. The structures are honest signs of the pelicans' superior fishing skills, but they are nevertheless burdens that are only useful for mating. As expected, the structures disappear when the mating season ends.[34]

Handicaps are often stereotyped as having no purpose other than serving as an honest and costly signal, but this is not always the case. Many handicaps still have some vital biological function. Let's use one of the most common handicaps—color ornament—as an illustration. Many animals, especially fishes such as guppies and sticklebacks, and birds such as finches and canaries, have bright coloration—red, orange, yellow—especially in males. How do bright colors evolve to be an honest signal while still serving some vital function?

Bright coloration in fishes and birds comes from carotenoids in the foods they eat.[35] Carotenoids are antioxidants that can enhance the immune system. As such, they are beneficial for the physical health of

FIGURE 4.3. (a) American white pelican with a "horn" on the top of its beak (photo credit: Len Blumin) and (b) great white pelican with a "bump" near the eyes (photo credit: Andrej Chudy) (CC BY-NC 2.0 licenses, no modification made to either photo).

animals, including humans. However, most animals can't synthesize carotenoids themselves. They must get them from the plants they eat. Therefore, bright colors serve as an advertisement of good health, good skills in finding quality food, or a good ability in converting nutrition—or a combination of these. Recent studies in birds have triangulated the genesis of bright coloration down to the expression of a cytochrome P450 gene: CYP2J19.[36] This indicates that males with bright colors may also have a high functionality of that gene. In this sense, carotenoid-related coloration is difficult to fake. This is another reason it serves as an honest signal of a high-quality male.

Moreover, bright colors also make animals stand out, attracting more attention from predators. Even for the human eye, it's hard to miss a scarlet tanager or an orange fish in the wild. Thus, bright coloration carries the burden of higher selective pressure from predators. If you can survive the onslaught of predation despite bold coloration, you are likely to have good genes. In other words, bright ornamentation is an honest signal for exactly the same reason it's a handicap. So, for females, there are multiple reasons for choosing males with bright colors.

Through these diverse and fascinating examples, we see how male animals use a range of tactics—some honest, some fake—to gain access to females for sex. In general, sexual selection via female mate choice forces males to show three types of adaptation: evolve honest handicaps, exploit female cognitive biases, or sneak for opportunities to mate. Therefore, sexual selection can result in honesty, fraud, and everything in between. This in turn leads to a diverse range of morphological and behavioral adaptations.

ﬓ

Although we've been using female mate choice to illustrate how females can enforce honesty in males by having a preference for handicaps, these same ideas can be applied equally to honest signaling in communication, in general, outside the context of mate choice. As long as a trait is difficult to fake, it can be used for signaling a variety of things, ranging from the readiness to fight to the willingness to collaborate.

Remember that whenever two animals cooperate, there is also a risk: if one cheats, the other incurs the costs. So, a good way to gain the benefit of cooperation while avoiding the risk of getting suckered is to talk with each other. When two parties can communicate, they can negotiate terms and assess each other's honesty. Because of this, ever since organisms began to interact, information warfare has existed. Let's look at some examples that demonstrate how handicaps are used in honest signaling.

In the great tit, an Old World bird related to chickadees, the black stripe in the middle of its chest is a status badge (fig. 4.4). The wider the stripe, the higher the degree of social dominance. Since these (male) birds often challenge each other, a subordinate can't wear a wider stripe to appear in a higher social level—in a challenge he will quickly be revealed as a fake. If your rank is actually lower than your stripe indicates, you will suffer more in challenges than if you wear a badge that suits your status. Here, scuffles between birds serve as a policing mechanism against fakes.

In the Gambel's quail, dominance is indicated by the crown feather. When it sticks forward, it signals "Don't mess with me!" or "Screw you!" But when it leans backward, it indicates submission—"Don't hurt me, please!"[37] The same can be said for a social wasp. Higher-ranking individuals have more and larger black spots on their heads, a badge that has withstood frequent tests by its rivals. If a cheater fakes its social status by having more black spots on its head, it will soon learn the consequence the hard way.[38]

In all these cases, the size of the status badge is tied to the level of physical challenge, with larger badges for higher levels, just like boxing matches categorized by weight classes. Here we can see, due to social enforcement, these status insignias are costly, forcing animals to honestly show their capacities. That is, they must back them up by being ready to be tested.

What we have seen so far are examples that show how individuals of the same species communicate honest information by using the handicap principle. This principle can also apply to honest communication between different species. For instance, many clients of cleaner fish use

FIGURE 4.4. A male great tit; notice the black badge in the middle of the belly (photo credit: Jiangxu Zhang).

a credit scoring system by assigning varying levels of credit to different cleaner fish. This allows clients to keep tabs on the quality of service so they can discriminate against cheaters, especially free riders.[39] Likewise, when being chased by a predator, many species of deer will flag their white rump patch, and antelopes will perform stotting behavior by leaping up. These both signal the same information to their pursuer—"I am very fit! Don't waste your time on me."[40]

꩜

Our own species employs honest signaling as well. Physical traits associated with female beauty—unblemished skin, lustrous hair, a perfect waist-to-hip ratio—are all signals of youth. Ultimately, as evolutionary psychologists have revealed to us, they translate into fertility, the all-important feature in the primitive eyes of men.[41]

Honest signals are far more prevalent in human cultures than some curious female body features. For example, many indigenous societies provide venues for open competition—such as singing, dancing, and wrestling—to give young people, especially women, a chance to choose their companions. The winners of such competitions are more desirable. These contests, like modern sports—many of which have their origins in such indigenous contests—allow honest displays of a person's skills, strengths, and capabilities, which are often rooted in the handicap principle.[42]

For instance, in Mali, the Dogon men dance with bulky masks during festivals. These masks can be so heavy that they have to be mounted onto the dancer's head by a couple of helpers. Burdened by the load, men dance with great caution, for their necks may be broken if they lose their footing and fall. So, only the strongest men can dance with the largest mask (fig. 4.5). Children, on the other hand, use small masks, presumably to get a taste of the adult world.[43]

In hunter-gatherer societies, men tend to go after large game animals. By doing so, they take higher risks, although the hunters don't necessarily contribute more calories than the gatherers.[44] When hunting generates less protein and calories than gathering does, it loses its primary function as a means of provisioning. Instead, it becomes more of an opportunity for men to advertise their skills, valor, intelligence, or invincibility—that is, it is used to send costly signals.

There's a reason that successful hunters bring back and share the spoils. They take advantage of the opportunity to show their largesse to their communities and gain social capital. Thus, sharing their spoils serves as a subtle handicap. Through this public relations campaign, they usually enjoy a better reputation, higher social status, and stronger political influence, all of which can lead to higher reproductive success in many indigenous societies, from the Ache of Paraguay to the Hadza of East Africa.[45]

Hunting large game is not the only way to show male prowess in small-scale societies. As an illustration, let's pay a visit to Mer (Murray) Island in the Torres Strait of Australia. The indigenous Meriam people grow yams as an essential staple food. However, men and women do it in very different ways. Women's primary concern is that their families have enough to eat, so they maximize the yield by digging shallow holes and growing as many yams as possible. Men are far less interested in efficiency and productivity. Rather, they grow yams primarily for social

FIGURE 4.5. A Dogon man dancing with a large mask
(photo credit: Erwin Bolwidt with a CC BY-NC
2.0 license, no modification made).

prestige, garnered from big-yam contests in the community. To prepare
for these events, they painstakingly dig large and deep holes and take
great care of their crops—but only to ensure that their yams grow as big
as possible. The problem is that this comes at a hefty expense in yield:
for every yam a man grows, a woman can grow 20. But men don't give
a damn, because their minds are preoccupied with the contest. If they
do win, they will become instantly well-known and respected. This will
in turn make them more likely to wield greater political power and social influence in their community.[46] So, for men, yams are not for the
table but are an honest display of their growing skills and an opportunity to gain social and political benefits.

I know village life quite well, for I spent most of my childhood at my grandma's village near the East China Sea. Among the approximately 200 villagers, everyone knew pretty much everything about everyone else. There was little privacy or secrecy, typical of a small, close-knit society. Whenever a man did something remarkable, such as hauling in a sea turtle or a big fish, killing a poisonous snake, or scaling to the top of a large tree, everybody in the village would soon know about it. The "oohs" and "aahs" that followed an outstanding feat couldn't be easily elicited from anything ordinary, such as a bucket of shellfish or a bushel of rice. This "Wow!" effect could lead to instant fame that might last far more than 15 minutes in the community. More importantly, it could generate considerable buzz among the villagers, especially women. When this happened, your fitness prospects would definitely be improved.

Costly signaling is no less common in industrial societies. You can see examples everywhere, such as luxury items like Ferraris, Louis Vuittons, and expensive digital gadgets of all kinds. These extravagant status symbols are at the core of what economist Thorstein Veblen called, over a century ago, conspicuous consumption (fig. 4.6).[47] Another example is partying college males who down bottles of alcohol to prove their physical prowess, especially when female peers are around.

Young men in particular are prone to engage in risky behaviors, openly performed in front of their peers: driving recklessly and partaking in risky sports such as car racing, skydiving, and bungee jumping. They want to show that they can come away unscathed, demonstrating their invincibility. The same principle works in the criminal world among gangsters because risk-taking can win respect from and acceptance by other members.[48]

In courtship situations, the handicap principle is readily apparent. Many men, for example, take on a big debt to buy a diamond ring for their fiancée or bride as a symbol of dedication—often because women demand them as proof of love and fidelity. But in reality, as a recent study shows, diamond rings symbolize the quality and worthiness of men as mates. Women would desire larger, more expensive diamonds when engaged to unattractive men, compared with attractive men.

FIGURE 4.6. Lamborghinis are commonly advertised for conspicuous consumption, often with a vulgar intent to pander to a primitive male instinct (photo credit: crguerra with a CC BY-NC 2.0 license, no modification made).

Apparently, good looks are a premium that can be compensated for with a larger handicap.[49]

Diamonds have little practical value except for cutting glass or making old-fashion mechanical watches, especially those that are vehicles of conspicuous consumption, such as a Rolex. Diamonds are not rare, either. It's been estimated that the diamonds in an asteroid crater in Siberia could be mined to alone supply the world for the next 3,000 years.[50] Also, because diamonds are extremely hard and difficult to cut, there is little room for artistic design beyond the stereotypical geometrical shapes. They can't be compared with glass or most metals in malleability for creative work. The only thing that makes diamonds stand out is their *perceived* value. Unfortunately, their market value is vastly inflated because of a monopoly by the De Beers Group, a multinational corporation that controls more than 80% of rough diamond distribution in the world.

Regardless of a diamond's intrinsic value, there are two primary reasons that diamonds have evolved to be a symbol of love in both Eastern and Western societies. One is that their high prices render them an ideal proxy of wealth. A one-carat diamond engagement ring costs $6,000, a considerable burden for the average American man. Such a gift would traditionally signify the ability of a man to earn enough to support a family. The other is that, unlike a BMW, which may be affordable for a short-term rental, a diamond ring is given forever to the fiancée. That's why, if the engagement is nullified, the ring must, in most cases, be returned. Once a diamond is given to a woman, the man is indentured by the financial burden. He is now bound to the relationship, willingly or not. This is equivalent to a bride price (given by the groom to the bride's family, paid either in valuable goods or through substantial labor services) still practiced in some traditional societies in Asia and Africa.

In modern nations with advanced economies and gender equality, vast numbers of women no longer count on men to support the family. As a result, diamonds have lost much of their luster as an honest signal of resource insurance. As their role of a costly handicap has eroded, the reason for their rise in value has started serving their downfall. It's foreseeable that the famous De Beers tagline—"A Diamond Is Forever"—is bound to be forgotten sometime in the future.

Some costly signals are quite subtle. For example, gifts are so commonly used in social exchanges that we tend to overlook the honest signals implied in them. An expensive gift is a costly signal, generally meaning a high level of sincerity. A trinket from a dollar store, on the other hand, makes you look stingy if you intend to show your gratitude to people who have done you a good turn.

Generosity can serve as a handicap that can bring significant social capital in the form of a good reputation. People often compete to show how generous they are.[51] Blood donation is a good example. Blood is popularly viewed as vital to one's health, even more so in Eastern than in Western societies. For this reason, blood donation is a costly signal to show one's good health and generosity at the same time—a twofold benefit to win respect from others in society.[52]

Costly signals are especially appropriate to demonstrate trust, solidarity, devotion, and faith. They can be expressed in many ways, ranging from enormous monuments, such as Stonehenge in England and stat-ues on Easter Island, to elaborate religious rituals, all carrying emotional or economic benefits.[53] On the surface, many religious practices seem enigmatic and maladaptive, including circumcision, fasting, and handling of dangerous animals such as poisonous snakes. But if we see these rituals as handicaps, they immediately make sense: they are ways to show devotion, build trust, forge loyalty, and facilitate cooperation, while deterring free riding among followers.[54] This provides a compel-ling explanation for an otherwise paradoxical phenomenon: the more onerous and restrictive religious practices are, the more donations and attendance they'll garner from church members.[55] The same logic can also explain why priests who practice Buddhism, Catholicism, and other religions willingly forgo reproduction to evince their piety to the divine. All these burdensome and elaborate rituals are handicaps serv-ing the same end: walking the walk, not just talking the talk.

By contrast, a signal is often ineffective at showing honesty when it can be cheaply reproduced. Politicians hold the Bible and swear to God all the time. These ritualistic maneuvers are not a handicap, because they're not costly. Politicians can make many nonbinding promises during campaigns but, once in office, they're free to forget they've ever made them and often end up serving themselves more than the people who elected them. One of the most common and egregious practices of politicians is the use of the "revolving door"—doing favors for cor-porate interests, then getting a well-paid sinecure when they leave office. That's why politician ranks among the least-trusted professions in America today.

As far as the handicap principle is concerned, public declarations by politicians are far less impressive than the ritual of *yubitsume*—cutting off one segment of your pinky to show that you are sorry in heart, not just in words. This ritual was invented and practiced by the Japanese underground organization the *Yakuza*. If politicians really want to gain public trust, they should try using a real handicap—such as a 10-year embargo on working for any of their corporate constituents after leaving office, which would apply to their close relatives too. Such a sensical policy could be implemented without difficulty if lawmakers had the will.

⌘

In this chapter, we started by examining the cost of honesty and then explored ways to offset this burden so that honesty can be a better policy than cheating. We've used the handicap principle to show how honesty can prevail in the context of sexual selection by mate choice. We then broadened the principle from gauging the quality of mates to assessing the level of honesty in all forms of communication in animal and human social interactions.

The three straightforward anti-cheating rules, the *Big Three*, based on evolutionary wisdom, are:

A. *Rely on traits that are hard to fake or cost a lot to have* (such as kinship, intelligence, and reputation),

B. *Impose handicaps to force displays of honesty* (such as celibacy, costly gifts, or hazing rituals in frat houses and military boot camps), and

C. *Police for rule compliance and punish rule violation.*

These can be readily used to promote honesty in our own species. While Rule A deals with naturally existing honest signals, Rules B and C show what we can do in principle to boost honesty by institutional means. They are aimed at either raising the cost or reducing the profitability of cheating, making the payoff of cheating less than that of honesty. Cost here could come in anything important for people such as money (as in the values of gifts), safety (as in hazing rituals), and reputation (as in sticking to celibacy).

The good news is that these rules are easy to follow, even in challenging situations. And we can derive inspiration from a far less brainy animal: the vampire bat. Despite its horrific reputation as a blood-sucking feeder, it's a paragon in building a prosocial society by using the Big Three.

Vampire bats will starve to death if they eat nothing for three days in a row. The challenge is that hunting for blood is anything but certain.

Many bats fail frequently in their hunting trips. This is particularly so for those less than two years old, as one-third of them fail on any given night. To deal with this problem, they've adopted a system of social insurance, in which blood is donated by those who are bellyful to those who are starving in the community. However, such a communal food sharing system is extremely vulnerable to collapse if it is overwhelmed by free riders. The bats avoid this problem by taking two measures:

1. Each community is made of genetically related individuals who know one another by individual-specific sound and possibly odor. Clearly, kinship is hard to fake. Thus, Rule A applies.
2. They also admit nonrelatives to the community but only those who reciprocate. As members keep tabs on others' reputations, free riders are ostracized and excluded from the community. Thus, Rule C applies.[56]

It is unknown whether Rule B is used to reinforce honest food sharing in the bat community. But there are signs that individuals showing largess in food sharing may be preferred as mates and rewarded with greater reciprocation, such as food or grooming services. We can see how the bats succeed in building a prosocial society by adopting at least two of the Big Three. They can serve as a model system for the triumph of honesty.

By now, we have explored how cheating is done and how honesty can survive and thrive amid pervasive cheating. As previewed in chapter 1, the next issue is the arms race between cheating and counter-cheating. What then will it lead us to? To answer this question, we turn to chapter 5.

CHAPTER 5

Catalyst for Innovation

As all parents can testify, raising children is tough. With your time, energy, and money all going to them, parenting feels like a self-imposed burden. It's best articulated in the lyrics of Sonya Spence's song, *No Charge*: "For the nine months I carried you . . . no charge / for . . . the costs through the years there's no charge."[1]

Even though you sacrifice years of your life, you are not necessarily rewarded for your efforts. Children who turn out to be ingrates or disappointments are by no means rare. Still, most people feel parenting worth the expense and trouble. Why do we have the urge to have children? Why would we pour all of our resources into raising a child, free of charge? The simple answer is that evolution has placed that obligation on us. Otherwise, the genes we inherited from our parents will come to an end with us. It would mark the demise of a genetic legacy that has defied astronomical odds to survive for billions of years.

The good news is that you don't have to forgo having children in order to enjoy a duty-free life. There is an easy way out: unload the burden of parenthood onto others. No money? No problem, especially when you can fly.

I am talking about parasitic birds—nest (or brood) parasites, as they're known in biology—who lay eggs in other birds' nests. Cuckoos are among the most notorious. Yet, contrary to their bad reputation for exploiting others' hard work, about 60% of cuckoo bird species hatch eggs and raise their own chicks. Unfortunately, the remaining 40% of cheating cuckoos are more than enough to tarnish the reputation of their species as a whole.

Remember the murderous cuckoo in the opening of chapter 1? Now, let's flesh out the story with the juicy details. During the breeding season, a female cuckoo hides herself near the nest of a host bird until she spots an opportunity. Next, she swoops down to the nest, throws out one of the host's eggs, and replaces it with her own.[2] She performs her work so efficiently that it takes no more than 2 minutes, sometimes only 10 seconds. (The record is 5 seconds, held by the bronzed cowbird.) Like shoplifting, speed and stealth are the keys to success when sneaking an egg into a host's nest. After the job is done, the cuckoo moves on, looking for another surrogate to raise her next chick.

Nest parasitism can be a highly profitable reproductive strategy. A common cuckoo female, for instance, can lay more than 25 eggs sequentially in a single breeding season, far exceeding the number a typical songbird mom can handle when raising her own brood. This would be impossible were the cuckoo unable to outsource childcare. How does she so successfully manage to get other birds to serve as her children's nannies, for free?

The short answer: she practices the Second Law of Cheating. She exploits the cognitive loopholes in her hosts, which include reed warblers, dunnocks, European robins, pied wagtails, and meadow pipits, all common in Eurasia. However, to get away with a scam of this magnitude, the cuckoo has invented a suite of bold and radical tricks. To follow the sequence of events, let's focus on the common cuckoo and the reed warbler to illustrate how they play the game.

We learned in previous chapters that evolution, as potent and inventive as it is, is unable to endow an organism with a survival skill that's irrelevant to its environment. Many birds are size-blind regarding their eggs because there was no selective pressure during their evolutionary history that made them care.

Niko Tinbergen, who won the Nobel Prize for his contributions in the study of animal behavior, once tested geese and plovers with false eggs of varying size, shape, color, and spot pattern. He was surprised to discover that these birds have only a vague idea about the appearance of their own eggs. As long as the shape is about right, they take them literally under their wings. Even more peculiar, they have an ingrained preference for eggs that are larger—often much larger—than their own. One of his students tried placing a volleyball in front of a mama goose sitting in her nest. It was supposed to be a prank, but the anecdote became legendary when the goose attempted to treat this gigantic round object as her own egg![3]

The goose is totally oblivious to the size of her eggs but dogmatically follows a simple rule: "Round objects in or near my nest are my eggs." In case you are quick to judge the goose as idiotic, this rule of thumb actually contains deep evolutionary wisdom that has worked well for the goose for eons. Because few objects in her world are egg-shaped, the rule virtually never fails. Thus, shape is all she needs to know about her eggs, and size is completely irrelevant. How likely is it that a volleyball will one day show up on the doorstep of her nest?

What works for the goose also works for the reed warbler, apparently. That's why the warbler follows a rule similar to the goose's rule. This mental glitch is exactly where the cuckoo chicks find their opportunity for a free lunch.

The evolutionary pathway for nest parasitism isn't a straight one, however. It winds and loops, steered by the process of frequency-dependent selection—a negative feedback loop between the parasitic cuckoo and the host reed warbler. For the warbler, if her nest has little chance of being targeted by the cuckoo, having a cognitive capacity sharper than needed would be overkill, wasting precious material and energy that could be put to better use. We learned how this worked in blind fish that live deep in caves in Mexico in chapter 3.

However, if the warbler's nest is parasitized often enough, having an acute cognitive system becomes a necessity, essential to protect her fitness by spotting and ejecting alien eggs from her nest. In this sense, the warbler is quite unlike the goose, whose maternal instinct is roused by any round object. The warbler, instead, has *some* capacity to discriminate between her own eggs and those of the cuckoo. This ability improves when she is victimized by cuckoos often enough.[4]

Evolution is usually a slow, incremental process, taking many generations to make something tangible happen. For the warbler under the threat of parasitism by the cuckoo, relying on natural selection to hone her cognitive capacity can't meet her immediate need. She has to act now with whatever she already has on hand. What can she do to put her limited cognitive ability to its best use in deciding whether she should accept or reject suspect eggs?

Simply, she can play probability to her advantage. In theory, if her nest is parasitized often enough, the likelihood of a false negative will be small. She should lean toward being trigger-happy and rejecting any suspicious egg she finds in her nest. On the other hand, if her nest is rarely parasitized, the probability of a false positive result will rise. Being trigger-happy may put her at higher risk of snubbing her own eggs—she might be throwing out her own baby with the bathwater, so to speak.[5]

This is precisely what reed warblers do in nature, based on a study by behavioral biologist Nick Davies and his colleagues. If their nests are parasitized 30% of the time by cuckoos, warblers are more likely to toss eggs out of their nests than when nest parasitism is rare. When the incident rate drops to 6%, however, they stop rejecting eggs—unless they find cuckoos loitering around their neighborhood.[6] Warblers, of course, know nothing about probability theory. Nevertheless, their cognitive capacity would enable them to unconsciously sense and track the risk, *as if* they knew the complex formula to calculate the odds mathematically.

Under pressure from the warbler's counter-cheating tactics, the cuckoo is forced to come up with new tricks to outwit her host. What can she do? The Second Law of Cheating points toward three possible paths to success. One is to produce eggs that are better mimics— beyond the capacity of the warbler to detect. This will certainly prompt the warbler to up her own game by improving her cognitive acuity, which will in turn spur the cuckoo to further polish her art in egg mimicry. So, for both parties, this evolutionary arms race is like playing a video game in which when you win at one level, you immediately face competition at the next level up (fig. 5.1).

The second path for the cuckoo is to find a new and less cognitively sophisticated species as her host. Such a move will generate an

FIGURE 5.1. Cuckoo eggs (identified by arrows) mimic the eggs of three different hosts in three European regions (Igic et al. 2012). *Left*: Hungary great reed warbler (*Acrocephalus arundinaceus*); *middle*: Finland common redstart (*Phoenicurus phoenicurus*); *right*: Czech Republic reed warbler (*Acrocephalus scirpaceus*).

evolutionary ripple in the local bird community, leading to strategic realignments for all involved. As the cuckoo expands her menu of potential hosts, it would seem to take her original victim—the reed warbler—off the hook. But this is not necessarily good news for the warbler. If she drops out of the arms race and loses her ability to distinguish cuckoo eggs from her own, she will remain as one of the cuckoo's main victims.

Nest parasites employ a range of evolutionary strategies to accomplish their goal. Some are specialists, taking advantage of a few selected hosts; others are generalists, laying their eggs in the nests of many species. Even though common cuckoos parasitize more than a dozen host types, each individual cuckoo tends to be focused on the ways she exploits her victims—such as mimicking the eggs of a certain species among all possible hosts. This strategic specialization allows individual cuckoos to solve two problems at once. One is to reduce competition from peers who are also seeking free nannies among a limited selection of hosts. The other is to avoid being a jack-of-all-trades, laying eggs to mimic those of many different species and run the risk of none of the tricks being good enough to fool their hosts.

Success in sneaking an egg into the reed warbler's nest is just the beginning. The cuckoo must continue to come up with tricks for every step down the line. It's like the commercial launch of a new gadget, such as a cell phone or gaming console. You can't simply place all your bets on a fast-processing chip. Everything else—hardware, software, peripherals, service—needs to be part of the package you're offering if you want your product to be viable in the market. Likewise, to make nest parasitism a success, the cuckoo needs a diverse set of strategies and tools.

After sneaking an egg into the warbler's nest, the next problem for the cuckoo is how to make her egg hatch first, ahead of the eggs laid by the host. If she fails, her chick wouldn't stand a chance against the warbler chicks in competition for food. To make the issue even more challenging, the cuckoo egg is larger and is expected to take more time to incubate than the warbler's eggs. How can the cuckoo enable her chick to succeed?

Two methods could help her achieve that goal. One, she can pick a warbler who is still laying eggs. This will buy her egg some vital time. Two, the cuckoo egg can speed up its development, faster than its size would indicate. These are precisely the methods the cuckoo employs. Some species of nest parasites (such as cuckoos and honeyguides) evolved a third, more reliable method: they can incubate the eggs inside their bodies to give them a head start before laying them in the nests of their hosts.[7]

Next comes another major challenge. Since the cuckoo chick is larger than the host chicks, and has a bigger appetite, how can it secure a sufficient food supply from its smaller surrogate mom? The solution: monopolize the food. What the cuckoo chick does, however, may be more than we can stomach. After hatching, it pushes every one of the host's eggs out of the nest so it can hog all the food brought back by the hardworking warbler. Cuckoo chicks have been known to commit this carnage while the warbler stands by watching in awe and utter bewilderment. Murderous behavior aside, the cuckoo chick becomes the image of cute eagerness, begging, chirping, and flapping its wings to motivate its foster mom to keep the food coming until it becomes strong enough to fly away—with nary a word of thanks for all the mom's efforts (fig. 5.2).

Finally, the third way for the cuckoo to outfox the warbler is to hack into a weak spot in the warbler's cognitive system that has little to do with its eyesight. Specifically, she can circumvent the warbler's frontline defense—the ability to visually distinguish her own eggs from counterfeits—by exploring loopholes in the warbler's learning process. This is like military deployment in the traditional battlefield. If your enemy's front is strong, you move to attack its undefended flanks.

FIGURE 5.2. A reed warbler feeding a cuckoo chick
(notice the size difference!) (copyright: Minden
Pictures).

Here's how the cuckoo does it. The reed warbler, like most other
birds, recognizes her chicks by a process called imprinting, a photo-
graphic type of irreversible learning, meaning that you can't unlearn
what you have learned, even if it's wrong. For instance, ducklings and
goslings can imprint on any moving object—be it a dog or a human—as
their mother.

Imprinting mostly works for baby animals, but in this case, it works
for new warbler moms as well. If a first-time warbler mom is imprinted
by a cuckoo chick in her nest, she will only take cuckoo chicks as her
children from then on, while rejecting her own chicks.[8] If she is hit with
this enormous cost, the warbler may be forced by natural selection to
find a new way to recognize her own chicks. But if such a disaster happens only rarely, the warbler population as a whole won't be hurt by
sticking to the same old rule: "Any chick in my nest is my baby." Thus,
the warbler may never have an opportunity to evolve the cognitive
capacity to tell her own chick from a cuckoo's.

🐦

Because Darwin's family wealth freed him from working for pay, he was
able to fully dedicate himself to science, producing 25 books over his

lifetime. Among them, the most quoted lines are in the last sentence of *On the Origin of Species*:

> There is grandeur in this view of life, with its several powers, having been originally breathed into a few forms or into one; and that, whilst this planet has gone cycling on according to the fixed law of gravity, from so simple a beginning endless forms most beautiful and most wonderful have been, and are being, evolved.

In poetic prose, this long sentence sums up how evolution creates the splendor of diversity and complexity. We have seen how much of life's diversity has evolved through paired arms races between predators and prey, parasites and hosts, germs and immune systems, and males and females. The pair of antagonists in the evolutionary game that has not yet received enough attention from us is cheating and counter-cheating.

How exactly can the interactions between cheating and counter-cheating produce diversity and complexity?

Cheating and counter-cheating, like all pairs of biological rivalries, enriches biodiversity through the never-ending cycle of an arms race, like that between a cat and a mouse. The need for sparring parties to outdo each other drives innovation in both. That is, cheating schemes spark countermoves, which in turn beget counter-counter-cheating maneuvers, ad infinitum. It's quite like playing chess. The invention of a new gambit will be met with a new countermove, and so on as the players each try to get the upper hand. In the process, a theoretically infinite number of tactics will be contrived over time. In biology, this process is manifested as new behavioral, physiological, morphological, and mental traits, all springing from cheating.

The common cuckoo gives only a keyhole view of how, from "so simple a beginning," cheating can blossom into "endless forms most beautiful and most wonderful." In fact, more than 120 bird species have evolved to be "professional" nest parasites. These are a grab bag of species from all over the world, including cowbirds, coots, parasitic finches, and honeyguides. Even the black-headed duck in South America uses this method to cheat.

In all species that have been shaped by cheating and counter-cheating arms races, such as the common cuckoo and the reed warbler, cascades of changes have been made, tweaked, and invented, first in behavioral strategies, then followed by other biological traits—physiology, morphology, life history. Although the specific biological traits involved can differ greatly across different parasites and hosts, the evolutionary motif looks familiar: convergence, in which like patterns emerge under like conditions. Or, revisiting the metaphor in chapter 3, the wheel can be reinvented on multiple occasions in different situations, but it meets the same need.

Following are a few notable variations on the theme of nest parasitism. These examples show behavior different from that observed in the common cuckoo. Couples of the great spotted cuckoo parasitize magpie nests and may play Bonnie and Clyde to pull off a major heist. If they target a particular magpie nest but the magpie couple are attentive, the male cuckoo will stage a mock attack to distract and draw the magpies away from the nest. This gives the female cuckoo the opportunity to sneak into the nest to get her job done.[9] In Africa, honeyguides lay eggs in the nest of several species of local birds, such as the little bee-eater.[10] However, honeyguide chicks are far more vicious than their common cuckoo counterparts. When the honeyguide chicks hatch, rather than pushing the host eggs out of the nest, they stab the eggs or chicks with a bill hook that evolved to serve this specific purpose alone (fig. 5.3). This enables the honeyguide chicks to monopolize food brought back by the hosts.

Parasitic chicks are born con artists. To satisfy their big appetites, they make louder begging calls to rouse their host's maternal instinct. Some species exploit the soft spot in their foster mom's sensory system by using morphological structures, such as bright yellow patches, to mimic the gaping mouths of fragile and needy chicks. For example, hawk cuckoo chicks in Japan have yellow patches on their inner wings. They flutter their wings and flash the yellow patches to urge their hosts to work harder as if constantly begging, "More, please!" The poor host mom sometimes appears confused by the wrong locations of the "mouths" and attempts to feed the yellow wing patches of the parasitic chicks.[11]

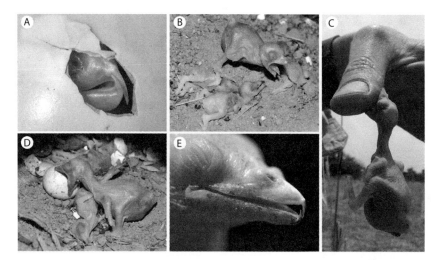

FIGURE 5.3. A honeyguide hatchling (a) bill hook; (b) with three bee-eater hatchlings it just killed; (c) biting a human hand; (d) biting a bee-eater egg; (e) about 8 days old (Spottiswoode and Koorevaar 2012, with a CC BY 4.0 license, no modification made).

To fight such maneuvers by nest parasites, some hosts evolved clever tactics with no need for sophisticated cognitive ability. Among such strategists is an Austrian bird called the superb fairy-wren. To distinguish her own chicks from those of their nemesis, the Horsfield's bronze-cuckoo, each fairy-wren mom sings a secret password to her nest before the chicks hatch. She finishes the education for her own chicks before the cuckoo chicks have time to learn. After hatching, if there's no password, there's no food. The parasitic chicks will starve to death.[12]

So far, all we have talked about are specialists in nest parasitism. How about generalists? How can they succeed? As a generalist nest parasite, you must adapt to a wide variety of hosts. Since most hosts are more or less capable of recognizing their own eggs, mimicking host eggs becomes increasingly difficult as more and more birds are added into the list of potential hosts—until mimicking hosts' eggs is no longer practical. How will you outwit the host's ability to detect fakes? The answer, unfortunately, is that you can't. You need a whole new approach to make a living as a parasite.

Now, think about looking for a job in Silicon Valley. You're surrounded by big tech companies such as Apple, Google, and Facebook. But you're a jack-of-all-trades. Although you know a bit of everything, you don't have the programming skills to become a developer. What can you do? One possibility is to become a manager. As a manager, your primary task is not to do programming yourself, but to supervise others' work. How can you do it? The answer, unsurprisingly, is by carrot and stick. While a benign boss may mostly use the carrot, a "bossy" boss may often resort to the stick. That's what generalist nest parasites often do. One master of this bullying tactic is the brown-headed cowbird.

The cowbird is a promiscuous nest parasite, in the sense that she lays eggs in the nests of over 200 bird species—pretty much any songbird who lives in their communities. How can she succeed in such a general approach? As we explained above, it's impossible to mimic all types of eggs in such a wide spectrum of potential hosts. So, the cowbird instead uses extortion to coerce her hosts into raising her chicks.

This is what the cowbird does. She first lays one or more eggs in the nest of a host and then waits around while keeping a watchful eye on the nest. If the host rejects the parasitic eggs, the cowbird will retaliate by destroying the nest structure, together with all the host's eggs or chicks.[13] She behaves like a street thug working for a Mafia boss. By ruining the host's reproductive stake, the cowbird leaves her victim only two choices, both of them bad: either comply by accepting the cowbird's eggs, which results in a partial loss of fitness, or reject the cowbird's eggs, which results in a complete loss of fitness. From an evolutionary standpoint of maximizing Darwinian fitness, the choice is obvious for the host. That's why the cowbird can make so many other species of birds do parental work for her.

The threat of destroying everything is the cowbird's stick; the carrot is the chance she offers her hosts to raise their own chicks as well, an incentive from the cowbird to make her hosts eager to do their work. (It's unclear whether this is the avian equivalent of Stockholm Syndrome, where hostages willingly cooperate with hostage-takers.) In any case, it's a good choice for the hosts in a bad situation. Indeed, birds who reject parasitic eggs raise 60% fewer chicks than those who kowtow to the bullies.[14]

Nest parasitism also occurs within the same species. In fact, within-species nest parasitism occurs in nearly twice as many birds—234 species and counting—as it does between different species. They include grebes, grouses, rails, coots, and many songbirds such as starlings, swallows, finches, and weaverbirds.[15] It's far easier to sneak an egg into a nest of your own species than that of a different species because you avoid the need for deception or the need to cover the trail. You can simply lay eggs in your own nest, then carry them to others' nests when unattended. This is exactly what some birds do.

Although nest parasitism is mostly studied in birds, it's also found in other animals such as amphibians, fishes, and insects.[16] It's especially common in social insects. For example, thousands of bee species lay eggs in others' nests. This occurs both within the same species and between different species. They are aptly named cuckoo bees.[17] Cuckoo bees most often choose closely related bee species as their hosts. Like their avian counterparts, cuckoo bees make their living by infiltrating their hosts' nests and devouring the young, while replacing them with their own eggs.

𝕩

Nest parasitism showcases how the evolutionary arms race between cheating and counter-cheating can lead to the emergence of complex biological properties in behavior, morphology, and life history, as we've seen so far. What else, you may wonder, may this arms race produce? One answer is social intelligence.

Before we get into the details of how this can happen, let's test ourselves with a simple puzzle known as the Wason selection task. (Please try this even if you've seen it before.) You are given four cards; each has a letter on one side and a number on the other. Your task is to answer the following question: Which card(s) do you have to turn over to check if the following rule is true: if a card has a D on one side, then it has a 3 on the other?

$$D \quad F \quad 3 \quad 7$$

The correct answer is cards D and 7. The other two cards are irrelevant. Puzzled? You're not alone. More than three in four people get it wrong. Even if you have seen this puzzle before, you may still fail. Why?

Before we answer, let's try a different version. You are in an American bar with an odd rule: customers are not allowed to talk with strangers, but each must present a card with truthful information about his or her age on one side and whatever he or she is drinking on the other. You see four young people sitting at a table, sipping their drinks and chatting joyfully. In front of them are the following cards:

$$\text{Beer} \quad \text{Coke} \quad 25 \quad 16$$

Which card(s) do you have to turn over to check for compliance with the drinking law (alcoholic beverages can only be consumed by people age 21 or older)?

The answer is "Beer" and "16." Easy, right? You're not alone here, either. Around three in four people get this one correct. You may have already noticed that the logical structure of this puzzle is exactly the same as the previous one. How come most people get this one correct, but fail the other? Apparently, context helps. The second one presents a familiar scenario of cheat detection, whereas the first one uses only abstract logic, with little practical relevance to our lives.[18] Fun exercise aside, this puzzle may hold the key to an important scientific question: How does intelligence evolve?

Intelligence—or more formally, cognitive ability—is often measured as the capacity for learning and problem solving in animals. One of its driving forces is an unpredictable environment, which renders it impossible for genetic preprogramming of strategies to cope. Cephalopods such as squids, cuttlefishes, and octopuses, for example, have to deal with catching their food while avoiding becoming food themselves. To thrive amid uncertain circumstances, they must adjust and adapt through learning. It's no wonder these animals exhibit a high level of intelligence supported by a complex nervous system.

Cephalopods would have been even smarter had they also been social. Changes in a social environment can take place at a moment's notice, much faster than those in the physical environment. Social animals are therefore doubly challenged by the need to adapt at the same time to the complexity of their peer environment and their physical surroundings.[19] As a result, two kinds of intelligence—individual and collective—are found in animals that live any kind of social life.

Collective intelligence, as demonstrated in such social (eusocial) insects as ants, bees, and wasps, evolved primarily in animals that are connected by kinship.[20] Their shared genetic interest smooths over conflicts of interest among peers, leading to a highly efficient, well-organized society, with all members serving in their capacities as workers, nurses, queens, drones, or soldiers. Because individual fitness is delegated to societal success, neither idiosyncrasy nor individual acumen is valued beyond the capacity for carrying out group projects. Therefore, it's no surprise that eusocial animals are known for collective intelligence but not for individual intelligence.

Individual intelligence (intelligence, hereafter) is favored in societies whose members have both shared and conflicting interests, depending on the circumstances. This is often seen in social birds and mammals. In such animals, society serves largely as a vehicle for its members to pursue their individual fitness, for which societal success is far less important than in eusocial animals. To get the most out of society, individuals need a powerful tool to evaluate, design, and execute strategies when dealing with peers. Here, the brain comes into play.

Social life is a double-edged sword. On one hand, you can promote your fitness by either cooperating with or manipulating your peers. On the other hand, you have to be careful, for you can fall victim to exploitation by your peers. That's why social intelligence is critical. It helps you find the best pathway through the jungle of convoluted relationships and develop a supportive social network. As a result, part of the primate brain evolved to meet the need of real-world tasks, such as distinguishing reliable friends from deceptive exploiters; less developed is the part of the brain needed to crack abstract math problems and solve logical puzzles, as demonstrated by the average person's performance on the Wason selection task.

Are animals with larger brains smarter? There's no simple answer to that. If it were true, whales would be far more intelligent than primates. Besides solving physical and social problems, the brain has many other things to take care of—adjusting hormone levels, sensing environmental conditions, controlling body movements. As in a jet plane, navigation is only one function among many and has little relevance to cockpit size. In primates, brain size is related to diet; fruit-eating species have larger brains than leaf-eating species.[21] Apparently, fruits are far less predictable and harder to find than leaves, so having a larger brain may help.

But finding food may require little social intelligence, as we have seen in cephalopods. Does social intelligence contribute to brain size? The answer is yes, but only for the part of the brain involved in high-level cognitive tasks: the neocortex in mammals. Interestingly, this part of the brain is related, first and foremost, to pair-bonding. Anyone who has been in a relationship knows it takes a lot of attention and effort to share your life with a partner. You need to communicate, coordinate, guess each other's intentions, design schemes to appease or trick your partner while at the same time figuring out your partner's ruses—to name just a few among many demanding social tasks. As we have seen repeatedly, regardless of cooperation or manipulation, society is an arena for realizing individual fitness.

If it takes a considerable amount of brain power to live with just one partner, how much more taxing is it on the brain to live with many peers? For primates, the answer is *a lot*. With more members to keep track of, animals living in larger societies tend to be socially smarter, if everything else is equal.[22] In primates, the relative size of the neocortex—that is, how large it is compared with the rest of the brain—rises with the size of the group they live in (fig. 5.4).[23] Apparently, as the social brain hypothesis suggests, the neocortex may have evolved mainly to deal with social issues and relationships.[24]

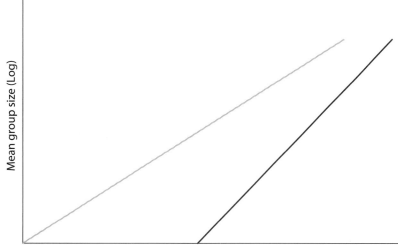

FIGURE 5.4. Neocortex size increases with social group size in monkeys
(gray line) and apes (black line) (redrawn from Dunbar and Shulz 2007).

Given the hypothesis that social living prompts the expansion of the
neocortex, you may wonder why species such as buffaloes and wilde-
beests that live in vast herds on the African savannas don't have an extraordinarily large neocortex relative to the size of their brain. There is
a catch to the hypothesis. Since these animals only come together temporarily for food, water, or safety, they never form a permanent group
where everyone knows everyone else. How about a deer or elephant seal
group with one dominant male herding females and young? There's still
a catch, but of another kind. The male only needs brute force to fight off
male rivals to protect his harem and young.[25] It doesn't require complex
cognitive skills to finesse their social environment. So, if pair-bonding
tells us anything about the evolution of the neocortex, it lies in *both how
often and how deeply an animal engages with its peers.* This unique aspect
of primate social life makes primates exceptional regarding brain evolution in general, and the neocortex in particular.

According to the social brain hypothesis, two main tasks your brain
undertakes are distinguishing collaborators from cheaters and, at the

same time, effectively manipulating others. For this reason, primatologist Richard Byrne calls social intelligence "Machiavellian intelligence." In a 1992 article, he and Andrew Whiten examined how often primates resort to tactical deceptions and found that macaques, baboons, and chimps stand out from the rest of the pack.[26] In 2004, Byrne assembled more evidence to show that deception is related to the evolution of the neocortex.[27]

To establish that Machiavellianism exists in primates, it's not enough to merely show that tactical deception is related to the neocortex. You need to show that social peers do in fact know what others think. That is, they can understand that others also have feelings, beliefs, and desires like their own. This mind-reading ability, known as a theory of mind (usually shortened to ToM) among psychologists, has been confirmed in chimps, bonobos, and orangutans in a recent study led by Christopher Krupenye.[28] In humans, telltale signs of mental perspective-taking may show up in children younger than one year. However, a full-fledged ToM capacity, including false beliefs, won't be obvious in children until they are older than four.[29]

ToM is important because it allows brainy primates to intentionally manipulate others. Here is an interesting account:

> The first time I saw an established leader lose face, the noise and passion of his reaction astonished me. Normally a dignified character, this alpha . . . became unrecognizable when confronted by a challenger who slapped his back during a passing charge. . . . The challenger barely stepped out of the way when the alpha countercharged. What to do now? In the midst of such a confrontation, the alpha would . . . writhe on the ground, scream pitifully, and wait to be comforted by the rest of the group. He acted much like a juvenile . . . being pushed away from his mother's breast. And like a juvenile, who during a noisy tantrum keeps an eye on Mom for signs of softening, the alpha took note of who approached him. When the group around him was big enough, he instantly regained courage. With his supporters in tow, he rekindled the confrontation with his rival.[30]

The above was an episode observed by primatologist Frans de Waal in a chimp group. Yet it feels so human that it reads almost like an ethnologist's field note. It's hard not to admire the way the alpha gained sympathy and recruited supporters through a carefully staged tantrum. How could he have concocted such a cunning strategy without high social intelligence?

The Machiavellian intelligence hypothesis, while certainly credible, may lean too much toward manipulating peers to outwit them in competitions. We're all aware that in human society, being deceptive, distrustful of others, and ignorant of social norms and ethical codes—all hallmarks of a Machiavellian personality—can be counterproductive, leading to loss of social support.[31] (In fact, Machiavellianism, together with narcissism and psychopathy, are known as the Dark Triad in personality research, indicating its undesirability as a trait in our society.)

Contrary to popular belief, people with a Machiavellian personality are not exceptionally intelligent, as shown by their mediocre performance in IQ tests. Nor are they better at mind games, as many may assume. They, instead, have a different mindset. When you propose a joint venture to them, their brains, instead of thinking, "Yeah, that's a good idea!" begin to plot and plan strategies to get more for themselves rather than being satisfied by taking their fair share. This reaction tends to chill their social-emotional response, making them appear to be cold and calculating.[32] At first glance, they may appear intelligent, charming, or even attractive. But in the long run, their tendency to be selfish, manipulative, and exploitative may backfire. In the end, they are usually no more successful than the average person.[33]

The rarity of people who practice Machiavellianism tells us that, like all strategies that rely on cheating, Machiavellianism is an alternative approach—a less-traveled path taken by only a few. This implies that cooperation, not manipulation, provides the main incentive to live socially. To put it in different terms, the primary theme for the evolution of social intelligence should be building alliances, including trust, reciprocation, compromise, peace-making, and other prosocial activities, to realize the inherent benefits of social living.[34] The primate social brain, accordingly, evolved primarily to iron out social wrinkles, to make the social environment a better place to promote the fitness of its members. If you envision social evolution as a dinner party, Machiavellianism is only an appetizer, not the main course, which is cooperation.

Byrne himself has recently broadened the context of the Machiavellian intelligence hypothesis and toned down the importance of tactical deception in the evolution of social intelligence:

> In giving advice to an aspiring prince, Machiavelli stressed the importance of being friendly, cooperative, kind, and generous—just until it paid not to be so. Similarly, the MI [Machiavellian intelligence] hypothesis applies just as much to the social sophistication shown in friendship formation, reconciliation, coalitions and alliances, kin support, and reciprocal altruism as it does to deception and overtly political manipulation.[35]

Having friends is always a good thing for the sake of cooperation. In fact, the more the better. However, the problem is that, as your circle of friends expands, so does the likelihood of including deceivers and manipulators among them. How can you maximize the benefits of cooperation while reducing the risk of getting cheated? One solution is to raise your social intelligence—sharpen your cognitive system so that you can more accurately distinguish who's honest and who's not. That's why we do better in detecting potential cheats when given a concrete scenario, as shown in the Wason selection task. Unsurprisingly, our memory can help us avoid being deceived, for we remember cheaters' faces far better than the faces of noncheaters.[36]

But there's a problem. Society can grow exponentially complex as the number of individuals (n) increases. The number of possible pair-wise interactions alone (without considering interactions of more than two individuals) rises as the function of $n(n-1)/2$, a function of near exponential growth. When you consider social networks of three or more people, the situation can be infinitely complex. But our social intelligence is not infinite, and our ability to spot lies is actually quite limited.[37] It seems that social intelligence may never be able to keep up as society increases in size. How can you cope? The answer: limit the number of your friends. You don't need lots of friends; you only need a stable circle of *close* friends you can trust. Quality trumps quantity.

How many close friends can one maintain stable, trusting, social relationships with? Before we can answer this question, we have to define who makes it into the circle of your close friends. According to anthropologist Robin Dunbar, they are "people you would not feel embarrassed about joining uninvited for a drink if you happened to bump into them in a bar."[38] Although amusing, this informal definition says a lot about who your close friends are. You probably know their ages, temperaments, and their relationships with others in your circle. You'd feel comfortable making off-color jokes or ribbing them playfully. Given these criteria, most casual acquaintances and professional contacts such as clients, students, and coworkers are out. Now, how many are left? Dunbar gives a crude estimate of between 100 and 230 with the mean about 150, give or take,[39] a figure that is now known as Dunbar's number.

Interestingly, Dunbar's number, though based on data from primates in general, appears to fit the size of typical groups in tribal and traditional human societies. In the Hutterites, for instance, a community may split into two when its members exceed this number. It's also close to the basic unit used by armies, both ancient and modern. Even communities as diverse as Facebook friends, terrorists, and cybercrime networks roughly follow Dunbar's number.[40]

Are these agreements just flukes? Hard to say. Rather, Dunbar's number may have some scientific merit, considering the push and pull of cooperation and deception within a society. It's easy to see that you can't be a bosom buddy with any stranger you meet. If you bet on the utopian ideal of "all for one and one for all," you may soon discover you're broke. So, you have to use your discretion, making sure your collaborators are not fake, pretending to helpfully cooperate, while in reality, taking advantage of your trust. To distinguish between the real cooperators and the fakes, you have to be able to know who they really are, what the content of their character is, who they like to hang out with, and whether you can count on them when their help is needed. If your circle of friends becomes too large, you simply don't have the time and opportunity to keep track, and to assess and update their personal information on a regular basis. Nor are you likely to have the time and resources to constantly reinforce your ties with them all. This puts a cap on the number of friends you can maintain close relationships with.

Social grooming is costly. Dunbar estimates that people in a cohesive group may have to spend nearly half their time getting to know other members and dealing with them in various ways. This leaves only half their total time for productive activities. If more is allocated to social grooming, they'll be even more strained for time. This sets a ceiling on the number of friends who can crash your beer party without feeling uncomfortable. So, 150 seems to be the upper limit of group size where members can maintain social intimacy. That's why tribal societies, subsistence villages, and military units all come close to this number.[41] Meanwhile, you should take with a grain of salt claims by those who boast of hundreds or even thousands of friends, physical or virtual. It's safe to assume that they don't know the difference between quality and quantity in friendship.

Be aware that social intelligence is different from nonsocial cognitive intelligence, also known as quantitative intelligence. The latter stresses problem solving, rather than reading the minds and assessing the behaviors of others. That's why we can be cognitively smart yet socially awkward—such as the stereotypical nerd. Conversely, people who are street-smart may do poorly in school. These stereotypes indicate that the two kinds of intelligence may have evolved rather separately. In a complex and competitive society like ours, however, those who are good at both have the best chance of success.

Fortunately, the two kinds of intelligence often go hand in hand. The problem is that intelligence may be used for the wrong purposes. Psychological studies show that cognitively smart children are more likely to lie because they are good at recognizing the opportunity and taking advantage of it.[42] At the same time, they also tend to be good at detecting lies so as to avoid being manipulated. While such a combination of social and cognitive intelligence rarely leads to criminal behavior in children, it may do so in adults. White collar crimes that involve cheating in business, finance, and government are overwhelmingly committed by (supposedly) smart and highly educated people, who can often avoid punishment. Did you ever wonder, for instance, how many big bankers were charged and convicted for massive corporate frauds committed by American financial firms during the real estate bubble in the 2000s?[43]

𝓕

Now, let's switch gears from social intelligence to mate preference. How does mate preference affect evolutionary trajectories propelled by cheating in general and manipulation in particular?

We've already discussed cognitive biases in previous chapters, especially in females. These biases are prone to exploitation by males, a common occurrence in sexual selection. Before the 1980s, however, researchers were unaware that such biases exist. Biologist Nancy Burley was among them.

In the 1980s, Burley was working in an aviary at the University of Illinois, investigating how zebra finch females choose mates. She was using the standard method of banding birds' legs with different colors to track their identities, behaviors, and reproductive histories. To her great surprise, this routine practice had a major impact on the birds' sex life. Female finches fell head over heels for males with red-banded legs, while rejecting those that were banded with green.[44] Apparently, red accentuated the orange color of males' legs, making them more attractive to the female eye. Why would female finches fall for such an odd artifact that doesn't exist in nature?

Alexandra Basolo, then at the University of Texas at Austin, was searching for the answer, not in birds but in small fishes. She noticed that male swordtail fish grow an elongated swordlike structure from the caudal fin (fig. 5.5). However, closely related platyfish, in the same genus *Xiphophorus* as the swordtail fish, have nothing like that. She wondered what would happen if a sword was added to a male platyfish who has no sword naturally?

She tried adding swords surgically. The effect on the platyfish was similar to the reaction of Burley's female zebra finches to males with red

FIGURE 5.5. Swordtail fish; note the male with a sword-like fin (photo credit: Gil Rosenthal's Lab).

leg bands: male platyfish, after the cosmetic perk, were suddenly more sexually appealing to females. Undoubtedly, female platyfish have a preexisting preference for swords, even though the male platyfish have no swords naturally. Apparently, male swordtail fish have evolved a way to exploit it, but male platyfish have not.[45]

Sensing a paradigm shift under way in the study of sexual selection, I made a scientific pilgrimage in the mid-1990s to the center of action, the Ryan lab in Austin, Texas, to visit Mike Ryan and Stan Rand. Unfortunately, by the time I arrived, they had just wrapped up an ingenious project. They used genetic markers (such as allozymes and mitochondrial DNA) to build an evolutionary tree that showed the branching pattern among eight extant species of frogs in the genus *Physalaemus*. Next, they inferred the mating call used by the ancestor of the two most closely related species by extracting common features in their calls.[46] Using this method step by step, they eventually figured out the call of the grand ancestor for all eight *Physalaemus* species. To test female response, they played back these calls to one of the living species, the túngara frog. As expected, females respond more enthusiastically to males of their own species than to those of a different species. Furthermore, females prefer male calls from species that are more closely related to them genetically, often *disregarding* whether they are current or ancestral relatives. Clearly, female preference for some features in male calls existed long before the túngara frog parted ways with its sister species.[47]

Although too late for the main course, I didn't entirely miss the dinner party. I found myself captivated by an ingenious experiment conducted by Gil Rosenthal, then a PhD student. Instead of the surgical procedure used by Basolo, Rosenthal let female fish watch what was jokingly called "fish porn"—video clips, in which male images had been manipulated, changing the length of the sexy sword. Indeed, females were more fascinated by virtual males with long swords than those with shorter or no swords, demonstrating again a preexisting bias in female preference.[48]

If one sparrow doesn't make a spring season, many do. The fascinating discoveries made in the Ryan lab were harbingers of the prevalence of preexisting cognitive biases, which have been found not only in vertebrates but also in invertebrates. In fiddler crabs and water mites, for instance, males have been found to take advantage of females by appealing to their sensory loopholes.[49] (You can find many more examples in Ryan's 2019 book, *A Taste for the Beautiful*.)

Unsurprisingly, sensory exploitation is common in primates, too. Many primates are known to have a penchant for red and orange, the colors of ripe fruits in the forests where they live.[50] For this reason, some species—most notably the bald uakari in the jungles of South America—bear these bright colors on their heads to attract females. These discoveries all point to the broad existence of hidden perceptual biases. The signs that were first glimpsed in Austin three decades ago have become established facts today.

While Austin researchers were celebrating their discoveries with music, beer, and Texas barbeque, Nancy Burley was quietly making a breakthrough in California. In 1998, a decade after her serendipitous discovery of the link between leg bands and sex lives in zebra finches, she and her collaborator Richard Symanski published a new study, in which they glued feathers onto the heads of male zebra finches and long-tailed finches (fig. 5.6). They found that females love males with

FIGURE 5.6. Male zebra finches (left three) naturally lack crest feathers. If you glue a feather, especially a white one, onto a male, females (right two) will show a more enthusiastic response (photo credit: Jim Bendon with a CC BY-SA 2.0 license, no modification made).

white feathers but abhor those with red or green. Since these finches have no natural crown feathers, how could females be drawn to a punkish-looking stud, simply because he bears such a thing?

The conclusion they came to seems obvious now: female finches have a hidden preference—a preexisting cognitive bias. But for Burley and Symanski, this wasn't enough. Females may have, as Darwin put it, "a taste for the beautiful."[51] At this point, disconnected information about sensory biases found in a wide range of animals began to come together to form a clearer picture, which foreshadowed an evolutionary explanation of something we humans have long taken pride in as existing in our own species alone: art.

🙉

Artistic creation relies heavily on illusions to generate desired effects by taking advantage of our cognitive biases and deficits. That is, it succeeds by using the Second Law of Cheating. In paintings, many objects, such as fire or sunlight, would be hard to represent on canvas without tricking our visual systems. Because we perceive things differently on different backgrounds, gray, for instance, can appear lighter when surrounded by white stripes than when they are surrounded by black stripes as shown in White's illusion (fig. 5.7).[52] And yellow, likewise, can brighten up.

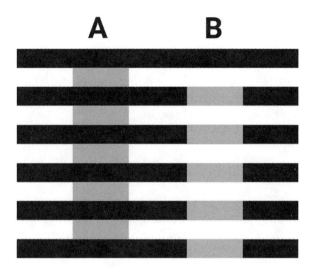

FIGURE 5.7. White's illusion; both gray columns (A and B) are the same color but appear different because of the color above and below the small gray rectangles (image from Wikipedia.org, public domain).

Rembrandt was a master in exploiting this optical illusion (color plate 11). His method has evolved to be a popular lighting technique in portrait photography, known aptly as Rembrandt lighting.

Visual illusions can be used in far more sophisticated ways. One such application was discovered by Durandeau, a Parisian with the keen mind of an entrepreneur. He observed that the appearance of a plain-looking woman was enhanced when she was seen with an ugly female companion. He was inspired by a brilliant idea: as Paris was never short of rich women who were anxious to be seen as more attractive, he could make serious money by running a rent-an-ugly-woman business. He did just that, and it didn't take long for his business to thrive.

Don't try a Google search for Durandeau's business, however, even though you may be curious to know more. This is a fictional plot in the 1866 short story, *Rentafoil*, by French writer Émile Zola. "Durandeau will be blessed by future generations," Zola concluded in the end, "because he created a market for a hitherto unsalable commodity and invented a fashion article which makes love easier."

You're not the first, Zola! Fish beat Durandeau by millions of years. Male guppies, for example, are known to pick less-attractive male peers with smaller color patches as their consorts, making them appear more attractive to females by contrast.[53] They use this deceptive trick to boost their value in the mating market.[54]

Impressed by the fishy trick? Wait until you see what Australian bowerbirds can do. Male bowerbirds have distinctly uninspiring plumage. But they make up for this shortcoming by building elaborate bowers, decorative structures that serve no function other than as enticements to impress females during the mating season. Females choose among a selection of males based on how well their bowers are constructed. To try to impress females and win a mate, males pour all their effort into building their bowers, including bonus perks. If females are attracted to red or blue objects, males will try to gather such colorful trinkets: flowers, fruits, or even pieces of plastic such as laundry clamps stolen from human residences nearby. They may steal ornaments from each other to spruce up their own bowers.

A recent study has revealed a truly amazing thing: in the great bowerbird, males can create a well-known artistic illusion called forced perspective to enhance the appeal of their bowers. This is a visual trick used in paintings, where smaller objects appear farther away, as seen in figure 5.8.[55]

If deceptive visual maneuvers in animal art seem primitive, they are used in much more methodical and sophisticated ways in human art. Painters have been systematically discovering and using optical illusions since the Renaissance to trick viewers into perceiving delicious fruits, open landscapes, or beautiful human bodies on a flat canvas.[56] Such illusions are even more important for modern paintings as they become abstract, increasingly removed from the real life content typical of classical paintings.[57]

Apparently, it all started in the mid-nineteenth century when photography was invented, posing an unprecedented challenge to realistic painting. It seemed only a matter of time before paintings would be superseded by cheaper and faster photographs. This perceived crisis drove artists to explore alternative modes of artistic expression beyond

FIGURE 5.8. Forced perspective: (a) an illusion in a bower; note that larger objects are strategically placed away from the entrance of the bower, whereas smaller objects are placed near or inside the entrance, creating the illusion that the bower appears bigger than it really is (photo credit: dracophylla with a CC BY-NC-SA 2.0 license, no modification made); and (b) a building (for comparison) where objects of the same size look smaller when they are more distant (photo credit: Lixing Sun).

realism. Among the pioneers were French impressionists, with Manet and Monet among the best known. Despite the obstinate rejection by advocates of mainstream neoclassicism (such as Ingres and David) and romanticism (such as Delacroix), impressionism triumphed. From that point forward, visual art evolved to be more and more detached from our perceived "objective" reality.

Is abstract art just another fad? Clearly not. Despite continual bashing by some traditional art critics and connoisseurs, abstract art has taken root and thrived as the new norm. And despite their modernity, abstract artworks fetch hefty prices in auctions, often higher than much older realistic paintings. (Think about the works of Munch, Picasso, or Pollock.) Why?

A main reason why abstract art excels lies in its unfettered capacity for expressing sensations. Also, since abstract paintings are not immediately comprehensible, viewers have to stretch their imagination to interpret them. This makes the creation and appreciation of art two separate subjective processes. Often, what the artists intend may be categorically different from what the viewers construe. Viewing art then becomes its own independent process, in which meaning is reinterpreted and sensation recreated. This makes the appreciation of art itself an art.

In my own experience, when I first saw Picasso's *Guernica*, I didn't understand what was in it, nor the story behind it, but I experienced a gut-churning sensation provoked by the imagery. This creative mental process gives viewers the opportunity to discover aspects of their own inner world, which may be unknown to even themselves. Picasso was well aware of this. "Painting isn't an aesthetic operation," he once said, "it's a form of magic designed as a mediator between this strange, hostile world and us, a way of seeing power by giving form to our terrors as well as to our desires."

Behind the curtain of artistic illusion is sensory exploitation. Unlike Nancy Burley's red-banded or white-crowned male finches that the females were strongly attracted to, many artistic creations don't meet with popular approval, but this doesn't deter artists from trying new things. They may eventually hit the jackpot, like the white crown in Burley's finches. But more often, they flunk. In this sense, artistic creation is a process of trial and error. A few artists get lucky, coming up with forms, styles, and genres that appeal to a wide audience, whereas the vast majority fail. Given this reality, many major art movements (with names that end with an intimidating -*ism*) may be little more than a post hoc confirmation of those styles that have met with popular acceptance.

Music works in much the same way. We sense sound by the organ of Corti, a thin strip of tissue that is curled inside the snail-like cochlea in our inner ear. The basal part of the organ is more sensitive to high-pitched sounds, and the apical part to the low-pitched sounds. So, keeping everything else equal (such as loudness, timbre, harmony, and rhythm), one way to enrich our music experience is to excite the entire perceptional range of the ear by producing all sounds from low pitch (such as cellos) to high pitch (such as violins). Such methodic exploitation of the human ear appears to be a reason why symphonic music became increasingly complex over time, evolving from ensembles of a few musicians in Bach's and Handel's compositions, to dozens in Beethoven's and Wagner's, to more than a hundred in compositions by Berlioz and Richard Strauss.

But the use of brute force to maximize our acoustic experience misses a major point: what we perceive as new, rich, and interesting is not simply a matter of greater quantity. Even though the entire range of the organ of Corti is saturated by a symphony, music cognition is still far more than sensing the sound at the peripheral level of the ear. It also takes place at two upper levels in the central nervous system: the lower limbic network and the higher left frontoparietal network.[58] Music can provoke our emotions because of cognition at these higher levels by at least six psychological processes: brain-stem reflexes, evaluative conditioning, emotional contagion, visual imagery, episodic memory, and the generation and violation of expectations.[59] That's why even musical elements—such as beat, pitch, timbre, tuning, harmony, and chord—as well as the holistic properties of an entire piece can generate rich aesthetic experience, which can inspire complex feelings such as melodious, calming, touching, beautiful, or peaceful sensations.[60]

Regardless of genre, such as classical, jazz, or pop, when we like a piece of music, our cortico-thalamo-striatal reward circuit lights up. If we don't like the music, our right amygdala and auditory cortex will intervene, making us ignore the music or turn it off.[61] That's why pop music, far simpler than classical, could still eke out a niche for its existence, especially in the twentieth century. Today, life without pop music is unthinkable for a vast number of people. Apparently, pop music has flourished by finding unexplored cognitive preferences in its audience.

Music cognition in the brain is poorly understood today. As a result, music creation remains a hit-or-miss process in searching for a sweet spot in a wide audience's preference, as it has always been. Even with the help of AI and Big Data, the success rate in finding the next big hit still falls woefully short.

It's clear that discoveries such as elongated swords in platyfish, forced perspective in bowerbirds, and crown feathers in zebra finches are not just peculiarities in lower and simpler animals. From them, we can trace the humble origins of our refined music and art, rooted in the discovery, through trial and error, of preexisting cognitive preferences.

𝄡

Ryan's sensory exploitation hypothesis assumes that sensory biases are static or evolve far more slowly than the traits that evolve to take advantage of it. This would give the traits enough time to experiment with variations until they discover what works best. This assumption is valid in most scenarios because traits used as signals generally evolve faster than preferences for the traits, as we have seen in insects, fishes, frogs, and primates. But there are exceptions for which the sensory exploitation hypothesis does not have a monopoly in the free market of scientific ideas.

As we've seen in the evolution of nest parasitism, when the nests of reed warblers suffer a high rate of parasitism, the warblers can evolve a sharper cognitive ability to counter deceptive maneuvers by cuckoos. Not all sensory biases stay stagnant for long periods of time; some can evolve quickly. In the context of sexual selection by mate choice, male traits used as signals and female preference responding to the male traits may tune to each other and evolve together, especially when prompted by changes in environment. This evolutionary pas de deux between traits and preferences is known as sensory drive, credited to biologist John Endler.[62]

This coevolutionary process can occur when a trait in males and the preference for the trait in females are genetically coupled. For instance, a gene coding for a larger trait in males is expressed in females as a stronger preference for the trait. This positive feedback between a male trait and female preference may lead to a runaway process, where the trait becomes increasingly exaggerated over time, resulting in such dramatic features as the extravagant tail of the peacock, the versatile song of the lyrebird, and the elaborate bower of the bowerbird, to name a few avian examples. This idea was first proposed by British evolutionary biologist Ronald A. Fisher.

The Fisherian runaway process is often seen as a competing alternative to the handicap hypothesis in the study of sexual selection.[63] They are not really that different.[64] The Fisherian process applies only to a situation in which a preference evolves quickly, whereas the handicap hypothesis works regardless of the rate of evolution of the preference.[65] For this reason, handicaps may have a broader explanatory power for sexual selection than the Fisherian runaway process.[66]

That said, we should never downplay the importance of the Fisherian process, since it's well-suited for both biological and cultural evolution. I learned this in 2013, when I came across an article in *The Economist* magazine, reporting that Bitcoin was trading above $300 apiece. Although the price seemed ridiculously high at the time, I still considered buying a couple of Bitcoins just for fun. But I decided against it when I realized that digital currency was often used for shady dealings. Additionally, it wasn't safe then to open an account with any of the exchanges (the Tokyo-based Mt. Gox, for example). What happened next was that the value of Bitcoin crept up slowly and steadily. Despite wild fluctuations at times, it eventually topped $64,000 apiece on April 14, 2021. This runaway process followed the exact evolutionary trajectory Fisher envisioned, if you simply replace the genetic link between a trait and a preference with the connection between rising price and a growing demand.

For any commodity, in general, its price will go up whenever there's a growing demand and soar when the supply doesn't catch up with the demand. That's why the price of 3M's N95 surgical masks went up manyfold for a short time during the early stage of the COVID-19 pandemic. This is a basic principle taught in ECON 101. But for behavioral economists, prices can also rise for two psychological reasons. One is the endowment effect: when we are selling something we own—be it a mug, a roll of toilet paper, or a share of Microsoft stock—we tend to ask for a higher price than we paid for it originally. The other is the marketing placebo effect, where our preference for something tends to increase with its price.[67]

To see this in action, put a $10 price tag on a bottle of wine and a $100 price tag on another bottle of the same kind of wine and offer them to your friends at a party. Your friends will inevitably tell you that the $100 bottle tastes better. If you can sell the $100 bottle for $200, the new owner is likely to prefer it even more, and so on. This is how the Fisherian runaway process goes. For Bitcoin, with only a tiny supply of about 20 million coins to meet a worldwide demand, plus the endowment and marketing placebo effects for the owners, it's no wonder that the price of Bitcoin could skyrocket.[68] In general, as long as there is positive feedback between an entity or a meme—a digital currency, a piece of art, a fashion type, an idea—and a preference for it, a Fisherian runaway process is likely to lead to a fad.

A primary difference between biological and cultural Fisherian processes is that cultural preferences change fast. Fads, by their very nature, come and go in quick succession. In financial markets, the mania surrounding the classic Dutch Tulip craze in the seventeenth century and the South Sea craze in the eighteenth century, to the Internet and real estate bubbles in the twenty-first century, all follow the Fisherian trajectory. We can usually find pump-and-dump fraud schemes behind such bubbles, as economist James Galbraith pointed out in his congressional testimony in 2010.[69] During the cycle, a lot of value, strategy, excitement, and ultimately destruction can occur in quick succession.

As we've seen, an element of deceptive manipulation virtually always exists behind a fad. It works particularly well for art. For example, on November 17, 2017, a small painting (65.6 cm × 45.4 cm; 25.8 in × 19.2 in) titled *Savior of the World* (*Salvator Mundi*) was auctioned off at Christie's for more than \$450 million, a new record for a painting at the time. Some 27,000 wealthy people—Alex Rodriguez and Leonardo DiCaprio included—showed interest, but most found themselves quickly outbid. Despite much ado, the auction was over in just 19 minutes.

What is truly amazing, though, is that the same painting was sold for a mere \$125 (in today's dollars) in 1958 and was described then as "a wreck, dark and gloomy." Then it was masterfully restored in 2010 by a world-renowned art conservator, restorer, and historian Dianne Modestini, who had spent five years on this project (color plate 12). The acceptance of the painting began to improve after a theory popped up out of the blue: it was Leonardo da Vinci's last work. Driven by the story, the demand began to grow, and the price climbed steadily, before skyrocketing. Even though sophisticated testing tends toward confirming its authenticity, it's still far from being proven. And some experts remain skeptical. But as long as the anonymous winning bidder believed the story,[70] little else mattered.[71]

Although the price for the *Savior of the World* seems outrageous, there is at least a *possibility* that the story of its creator is true. Oftentimes, authenticity is unnecessary for an artwork to soar in value as long as the demand is rising. In 1961, Piero Manzoni, a renowned Italian conceptual artist, created 90 tin cans of readymade artwork under the collective name *Merda d'artista*. These creations were sought after by museums and private collectors. In 2002, the prestigious Tate Gallery in London acquired one of them, Can 004, for \$61,000, adding it to several others already in its collection. In reality, however, these cans were fraudulent. The artist-cum-prankster made no attempt to hide the fact, since he labeled all the cans in both Italian and English—*Merda d'artista*, or Artist's Shit, with an explicit description, "Contents 30 gr net. Freshly preserved, produced and tinned in May 1961."

Manzoni did it as "an act of defiant mockery of the art world, artists, and art criticism." Even so, a spokeswoman for the Tate defended the museum's action, claiming that Manzoni "was an incredibly important international artist," and the acquisition of Can 004 was a "very important purchase for a very small amount of money." She was probably right if we consider how lucrative art dealing can be—despite the irony behind Manzoni's sham artworks made of his own digestive waste. Today, *Merda d'artista* are still in demand among the artist's collectors. Since these tin cans were not properly autoclaved to sterilize the contents, more than half of them leaked and rotted over time, making those that remain even pricier.[72]

In both of these cases, we can see that as long as there is rising demand for a piece of art, its price can go crazy through a runaway positive feedback process. This can happen regardless of whether it's an unverified masterwork or a can of human excrement.

𝓴

In this chapter, we have shown how cheating can serve as a catalyst to trigger a cascade of evolutionary changes and innovations, leading to new traits in behavior, physiology, morphology, life history, and even beauty in the biological world. Specifically, the perennial arms race between cheating and counter-cheating can spur the emergence of such complex properties as social intelligence and art in animals and humans. Without the catalytic power of cheating, our world might be quite boring and without a wide variety of biological and cultural diversity.

If you have mixed feelings about cheating, I'm with you. We need to take a fresh philosophical perspective on the subject to see whether certain kinds of cheating may be considered legitimate. But before we delve into this, we need a better understanding of cheating and self-deception in our own species.

CHAPTER 6

Cheating in Humans

In the summer of 1971, a team of eight American flight attendants, led by a young pilot, all in impeccable blue Pan Am uniforms, were touring European cities—London, Paris, Madrid, Rome. Their mission: promote the image of the airline, the most prestigious carrier in the world at the time. Air travel was fast, comfortable, and elegant, with travelers who were well-dressed and groomed. It was a desirable aspect of lifestyle, especially for the well-off. The Pan Am crew always made a splash as they paraded through the ritziest parts of the town.

The entire tour was a gigantic scam, however, conceived and orchestrated by the "pilot" Frank Abagnale Jr., then only 23 years old. The "flight attendants" were college students he'd handpicked from hundreds of applicants from the University of Arizona. None of them had anything to do with Pan Am. Abagnale hit the jackpot: not only was the two-month grand tour of Europe free for both him and his glamorous cohort, but it also lined his pockets with $300,000 (millions in today's dollars). Staged openly and blatantly in the busiest streets of European cities, the scam may have been the most daring the world has seen in recent history.

If you have read Abagnale's autobiography, *Catch Me if You Can* (or watched the movie by the same name),[1] it's hard not to marvel at his ability to pull off his audacious stunts. Abagnale writes that he

> rips off several hundred banks, hustles half the hotels in the world for everything but the sheets, screws every airline in the skies, including

most of their stewardesses, passes enough bad checks to paper the walls of the Pentagon, runs his own goddamned colleges and universities, make half of the cops in twenty countries look like dumbasses while he's stealing over 2 million.

How could Abagnale succeed in pulling off such bold and brazen frauds? This is the question we'll attempt to answer in this chapter.

Cheating in humans is unrivaled in the animal world—whether in scale, variety, intricacy, or novelty. This is due primarily to three factors: the use of language, a high level of intelligence, and the complexity of human societies. Language provides a new powerful tool to lie and deceive; intelligence facilitates the invention and design of schemes; and societal complexity supplies a wellspring of opportunities to defraud. For these reasons, a book about cheating that did not include a discussion of the issue in our own species would feel like a drama without a climax.

As we try to uncover Abagnale's modus operandi, we'll search for answers to three questions about human cheating: What do cheaters cheat for? How do their schemes work? And whom do they prey on? In doing so, we hope to find patterns in human cheating and figure out how they fit into the big picture of cheating in the biological world.

🜊

If there were an award for accomplishments in cheating, Abagnale would be a serious contender for top prize. Although a high school dropout, he succeeded in impersonating a Pan Am copilot, a California pediatrician, a Louisiana lawyer, and a Brigham Young University professor, to name just a few. Even more amazingly, he pulled off these escapades primarily in his teenage years, between the ages of 15 and 21. When he was finally caught in Europe and extradited to the United States, he made a miraculous escape from a lavatory soon after his plane landed at JFK airport in New York City. Even after he was recaptured, convicted, and sent to prison, he walked out of the penitentiary by making the jailers believe he was an undercover FBI agent.

This brings up our first question: What do human cheaters cheat for? The answer, based on Abagnale's exploits, comes down to one word: resources. Biologists sort resources between two bins: those that can help survival (such as food and shelter) and those that can boost reproduction (such as mating opportunities and quality of mates). Both can be ultimately boiled down to one quality meaningful in evolution: Darwinian fitness. In this context, human cheating is both universal and unique. It is universal because human cheating is driven by the same biological instincts and uses the same two rules shared with other animals. It is unique because human cheating embraces the level of diversity, complexity, and ingenuity that is far beyond the reach of any other animal. And it keeps pace with changes in our culture, especially social institutions and technological innovations.

In our everyday lives, a good part of cheating is for material resources— money in particular.[2] Abagnale cheated by creating false personas of various kinds in regard to his profession, skill set, educational background, and all other details essential for the success of his schemes.[3] Disguised by false identities, he forged checks and cashed out a total of $2.5 million in the 1960s. This was a time when millionaires were few and far between in America, and he had far surpassed that milestone before reaching the legal drinking age.

Cheating for money comes in a variety of forms. These include scalping tickets, dishonest business practices, embezzling funds from organizations, cooking up Wall Street Ponzi schemes, and hiding money in shell companies. Lawyers and consultants, for instance, overcharge clients; physicians and dentists overtreat patients; citizens evade taxes by claiming dead family members as dependents; fossil-fuel companies refute global warming by denying scientific facts; and politicians . . . well, we don't need any elaboration of the myriad ways politicians cheat.

Beyond money, people also cheat for nonmaterial resources such as status, reputation, social credentials, and career opportunities—which can all be converted into material gains down the line. For example, college education has no material value per se, but it's a conduit to career success and future earning capacity. That's why some parents "help" their children get into prestigious schools by forging test scores, faking extracurricular activities, and/or bribing admission officers and varsity sports coaches—as a recent FBI sting that exposed cheating in college admission illustrates. It's also why 68% of college students cheat on tests and assignments at least once. Similarly, students claim sudden deaths in the family, mostly grandparents, a week before a major exam. Curiously, the rate of such incidents increases linearly with decreasing grades.[4]

People use a range of dishonest tactics to gain recognition and promote their social status. These include self-promotion, taking credit from others, and fawning over higher-ups. And we tell white lies almost daily at the workplace or among friends and relatives to build better relationships. These are small transgressions most of us commit to increase our social capital, which can later be translated to material gains. Billy Joel is not exaggerating reality with his lyrics, "Everyone is so untrue."

For Abagnale, cheating for money was a pathway to an even more basic biological need: sex. At the age of 15, he needed money to go on dates, but he had no special skills. Manual labor didn't pay much, so he began to resort to minor frauds to earn more. That started him down a slippery slope. He soon upped the scale of his operation, wading deeper and deeper into the treacherous water of deceit. By his twenty-first birthday, his check-writing frauds had already claimed victims in all 50 US states and 26 nations around the world.

Abagnale was by no means unique in this regard. The market for gaining a sexual relationship is riddled with dishonesty. Among the most common are little lies used to attract mates. For example, men seeking dates online will try to enhance their profile by exaggerating their income level and adding several inches to their height. Women, meanwhile, lie about their body weight, understating it by 15 pounds on average.[5] So, if you meet somebody online, be ready for surprises on your first face-to-face date.

Married people cheat too. Infidelity among humans is so common that the very word "cheat" carries the connotation of "a sex cheat" in our everyday parlance. Data show that about 1.5–5% of people engage in extramarital sex per year.[6] This is not a negligible rate, for it can add up to 22–25% for men and 11–15% for women during a lifetime.[7] Note that

PLATE 1. A hawkmoth caterpillar mimicking the head of a viper (photo credit: Andreas Kay, In Memoriam: Ecuador Megadiverso (webpage) with a CC BY-NC-SA 2.0 license, no modification made).

PLATE 2. A wasp "mating" with an insect-mimicking orchid flower (photo credit: wislonhk with a CC BY-NC 2.0 license, no modification made).

PLATE 3. Type I male (center left), female (center right), and type II males (on the sides) in plainfin midshipman (photo credit: Andrew Howard Bass).

PLATE 4. Three types of male side-blotched lizards: (a) orange, (b) blue, and (c) yellow (photo credit: (a) and (c): tombenson76 and (b): Ranger Robb with CC BY-NC-SA 2.0 licenses, no modification made to any photo).

PLATE 5. The gray and green morphs of Pacific treefrogs (photo credit: Lixing Sun).

PLATE 6. The rotating snake illusion showing the effect of motion dazzle (image credit: Dennis S. Hurd with a CC BY 2.0 license, no modification made).

PLATE 7. Warning colors in (A) harlequin bug, (B) eastern newt, (C) burnet moths, (D) sea slug, (E) eastern coral snake, (F) Appalachian mountains millipede, (G) harlequin poison frog, and (H) ladybirds (Rojas et al. 2018).

PLATE 8. Müllerian mimicry in Heliconius butterflies (from Meyer 2006).

PLATE 9. King snake and coral snake. Can you tell which is which? (photo credit: (a) 2ndPeter and (b) Beaver w/ a Toothbrush with CC BY-NC-SA 2.0 licenses, no modification made to either photo).

PLATE 10. Birds of paradise evolved a stunning diversity of elaborate ornaments in males as handicaps (illustration by Szabolcs Kokayart, from Pratt, T, K., Beehler, B. M., and Kokay, S. (2014). Birds of New Guinea. 2nd ed. Princeton: Princeton University Press).

PLATE 11. *Philosopher in Meditation* by Rembrandt (photo credit: jean louis mazieres with a CC BY-NC-SA 2.0 license, no modification made).

PLATE 12. *Savior of the World* (supposedly by Leonardo da Vinci), before (center) and after (sides) restoration (photo credit: Daniel Arrhakis with a CC BY-NC 2.0 license, no modification made).

these numbers are conservative estimates based on "honest" confessions by cheaters. The real scale of infidelity is likely far more substantial than this.

While some see cheating as a disgrace, others see it as an opportunity for profit. One such business adventurer is Ashley Madison, a Canadian company that provides online dating services for married people. Its banner slogan—"Life is short. Have an affair."—sounds anything but apologetic for promoting infidelity. Although the company now boasts 60 million people in its registry, a hack in 2015 revealed that their database was overwhelmingly men. Jeff Ashton (Florida State Attorney), Jason Doré (a Louisiana GOP official), and Josh Duggar (a small-time star in the reality show *19 Kids and Counting*) were among the notable users.[8] Women, it turned out, were in short supply. To satisfy lopsided male demand, the company used bots to impersonate women. Thus, the cheaters themselves became the cheated. And the business of Ashley Madison is of cheaters, by cheaters, and for cheaters.

If the gender imbalance at Ashley Madison tells us anything, it's that men, like most male animals, have a stronger impetus to cheat. This has been borne out by data.[9] That's why Jackie Kennedy Onassis stated ruefully, "I don't think there are any men who are faithful to their wives." Nevertheless, we shouldn't downplay the scale of marital cheating among women, who tend to do it in subtler ways.[10] Female desire is stronger around the time of ovulation, a fact made clear by women's self-confessions.[11] It may be related to high estradiol and low progesterone levels around this time.[12] Interestingly, women in ovulation are far less likely to use contraceptives, indicating that infidelity may be for the purpose of out-of-wedlock conception.[13]

Many biological and cultural factors—such as hormones, genetic makeup, intelligence, and tradition—affect cheating for sex. There is even a role played by religiosity. Statistics show that extremely religious people and unreligious people are more likely to cheat than those who are moderately religious.[14] The reason behind this inverted bell curve of infidelity, however, remains unknown at this point.

Genes are also implicated in cheating for sex. It's known that the genes coding for vasopressin and oxytocin receptors in the mammalian brain affect the quality of pair-bonding, and this in turn leads to more or less sexual infidelity.[15] A recent study attributes women's interest in extra-pair affairs to certain variants of the vasopressin receptor gene. Although there are many other factors involved, this genetic variation alone accounts for 40% of female promiscuity.[16] Another variation that affects both men and women is found in the dopamine D4 receptor gene. One variant may turn people into thrill seekers, looking for novelty to make life—including sex life—more exciting. As a result, the rate of infidelity and promiscuity can rise by as much as 50% in people bearing this version of the gene, compared with the general population.[17]

When we look at all the variations of human infidelity, we find a familiar pattern shared with other animals: men typically cheat to expand their reproductive opportunities, whereas women do it to gain access to resources and/or better genes for their children.[18] Therefore, human infidelity, although diverse and complex, still falls within the boundaries of evolutionary theory, most notably Bateman's rule.

𝔑

As we know from the previous chapter, in socially monogamous animals such as most birds and a few mammals, females may cheat on their male partners for their own sake. Extra-pair paternity—or "cuckoldry," a word aptly derived from the cuckoo—is a major cost to male fitness, especially when males invest heavily in caring for and raising offspring. To counter double-dealing, many male animals resort to mate guarding to guarantee their genetic heritage.

In humans, extra-pair paternity without the father's knowledge is around 1% and 1.7% per generation in traditional societies[19] and in modern Western nations,[20] respectively. But these small mean numbers should not mask significant variations in different situations. In some cultures, extra-pair paternity could be much more substantial. In the Himba of Namibia, for example, it could be as high as 17%.[21] (This is not to mention that mate poaching—a paired mate lured by an outsider into a new relationship—is prevalent and persistent.[22]) Therefore, mate guarding pays off.

The basic forms of human mate guarding are similar to those in many other animals: divorce, desertion, and violence. While women often leave men who cheat,[23] men are more likely to use physical force, manifesting the biological legacy of the Stone-Age brute that still resides within us today. Female infidelity is one of the most common reasons for male domestic violence including beating, raping, or killing their wives.[24] It's distressful to know that rape and honor-killing, although rare now, are still practiced in some cultures to discourage women from engaging in extra-pair affairs.

Threats and retaliations against cuckoldry are only means to the end of preventing extra-pair cheating from taking place. A more proactive alternative is to control women's sexuality. In a patriarchal society, men can do three things in this regard: conceal women's attractiveness, restrict women's mobility, and reduce women's sexual desire.[25] These measures are only variations serving the Stone-Age purpose of mate guarding and are not unique to humans. As we saw earlier, male garter snakes are known to perfume their serpentine mates with the chemical squalene to render them unattractive to rival males. But with our unique linguistic capacity, high intelligence, and societal complexity, human mate guarding takes on a far more "creative" bent. This is where a wide and elaborate set of cultural practices play out.

In some traditional societies, for example, men stay at home during their wives' most fertile periods. Guessing when that is can be problematic at best, since the specific time of ovulation is mostly unknown even to the women themselves. For this reason, men often resort to more reliable methods. The chastity belt was invented for this purpose in medieval Europe for times when men were away.[26] Many Eastern cultures came up with a different solution: cut the wives off from any contact with other men by keeping them deep in the house. That's also why, in Japanese, "inner person" is a modest way to refer to a wife. The Chinese did even more than that. For centuries, from the Song (960–1279) to the Qing Dynasty (1644–1911), a massive cultural campaign was waged to promote "lotus feet," forcing young girls to arrest the development of their feet by foot-binding (fig. 6.1). As a result, women's mobility was greatly reduced.[27]

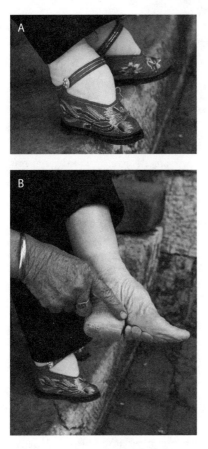

FIGURE 6.1. A woman with "lotus feet" (a) with 4-inch shoes, and (b) the deformed feet (photo credit: Dr. John Bullas with CC BY-NC-ND 2.0 licenses for both photos).

But don't laugh at such byzantine cultural practices. There are odd laws that still exist in the United States today that had their origins in the male instinct to guard his mate. For example, women are not allowed to enter Congress wearing sleeveless tops or dresses. They're not allowed to wear pants in Tucson, Arizona, or expose too much cleavage in Cleveland, Ohio.[28] Although these obscure laws are rarely if ever enforced, they remind us of the time when men ruled.

Patriarchy is on the decline in the Western world, but it is still alive and well in many other regions. Although Saudi Arabia was applauded for recently lifting its ban on female drivers, women in several Arab nations are still not permitted to go to a stadium to watch a soccer game or travel unaccompanied by male relatives. In some Muslim countries, they are still required to wear coverings such as the burqa, hijab, or chador. The biological meaning of this cultural practice has been systematically investigated by a group of Iranian researchers led by Farid Pazhoohi. Their studies confirm the popular belief that by covering facial features and body curves, veiling reduces women's sexual appeal and prevents them from making eye contact with potential mate poachers.[29]

Probably the most controversial practice tied to male mate guarding is genital mutilation, including excision, infibulation, and clitoridectomy.

As many as 130 million women have gone through these painful and unsafe procedures even in modern times, mainly in 29 African nations. The purpose of genital mutilation is to restrict or control women's sexual attitude, desire, and behavior so as to cut down the risk of extra-marital affairs.[30]

ℛ

How do cheaters cheat? We can find many of the answers embedded in the adventures of Frank Abagnale. It was his "street smarts" that made him so incredibly successful. He did not operate in disreputable back alleys but on Main Street, where honest and decent people bustle. His talent lay in his ability to observe and mimic behavior. He turned himself into such a superb human chameleon that he could don or doff bogus identities at will. This social camouflaging ability was key to the success of his schemes because it enabled him to gain others' trust.

Despite their high level of sophistication, Abagnale's tactics fell squarely within the two primary methods used by animal con artists: abusing the honest elements in communication (First Law) and exploiting loopholes in his victims' cognitive systems (Second Law). It is through the use of the First and Second laws that all human cheaters conceive, design, and execute their fraudulent schemes.

This is a good time to bring up the subtle distinction between lying and deceiving in human cheating. *Lying is sending false information, the domain of the First Law; deceiving is exploiting the biases, weaknesses, or deficits in the cognitive systems of victims, the domain of the Second Law.* In many schemes, however, both laws apply. For instance, flattery is to falsify the message (First Law) as well as pander to the cognitive biases in people who generally prefer to hear nice things about themselves (Second Law).[31] With this distinction in mind, we'll continue to use the word "cheating" to mean both lying and deceiving.

Here is how Abagnale skillfully applied the two laws of cheating. Always a keen observer, he launched his career as a master fraudster based on two insights of human nature. First, he noticed that people naturally had a high level of trust in those who are in respected professions. It was, and still is, a common cognitive bias: judging strangers by gut feelings, based on what psychologists call System I thinking—thinking quickly without thinking critically. He also observed that when cashing personal checks, cashiers would normally avoid asking for more than an ID card, which could be easily forged.[32] He posed as a pilot by simply wearing a pilot uniform he got from a special supplier. "No one checked with the bank to see if the check was good," he chuckled to himself, after his first successful forgery attempt.

He improved the efficacy of his scams over time by further exploiting people's cognitive biases. He had his fake checks made by professionals with special printing equipment and material. As a result, his forged checks looked so real that they passed inspection with the naked eye. That attention to detail enabled him to pass bad checks again and again without being caught.

Abagnale also methodically took advantage of loopholes in the banking system. He discovered that it took around five days for a bad check to bounce. This delay provided him with plenty of time to vanish safely from the scene. When suspected or questioned, he was skilled at diverting attention from the issue at hand by providing false information that couldn't be immediately verified. This is exactly why he could get away with taking a team of college women on a European tour. Who on earth would think such a large outing could be faked? Even for those who might question it, how would they know whom to contact in the labyrinthine bureaucracy of a large company? On one occasion, when his fraud was about to be exposed, he slipped away by disguising himself as an FBI agent. Who would mess with an FBI agent by questioning his authenticity? These clever maneuvers disarmed the defenses of those who had doubts and misgivings until it was too late.

Are you wondering how a teenaged Abagnale could be so good at impersonating professionals whose jobs required years of training? The simple answer is that people tend to judge others by their appearance. Abagnale looked much older than his age. To capitalize on this natural advantage, he added ten years in his forged IDs, moving his birthday from 1948 to 1938. Now, his fake age and unusually mature appearance matched, adding credibility to his deceptions.

To make his charades even more convincing, he did his homework. He steeped himself in the knowledge and lingo of whatever profession he pretended to be in. To masquerade as a pilot, he "started haunting the public library and canvassing bookstores, studying all the material available on pilots, flying and airlines." To feign a physician, he "quickly acquired a broad general knowledge of pediatrics, enough . . . to cope with any casual conversations concerning pediatrics." To pretend to be a sociology professor, he audited a sociology class before he taught it. In all these cases, his superficial knowledge was sufficient to fool even the professional gatekeepers. No wonder it was easy to pull the wool over the eyes of lay people.

He writes that his modus operandi consisted of three factors:

> The first is personality. . . . Top con artists . . . are well dressed and exude an air of confidence and authority. They're usually, too, as charming, courteous and seemingly sincere as a politician seeking reelection. . . . The second is observation . . . with the ability to pick up on details and items the average man overlooks. The third factor is research. . . . A con artist's only weapon is his brain. A con man who decides to hit the same bank with a fictitious check or a sophisticated check swindle researches every facet of the caper. . . . I knew as much about checks as any teller employed in any bank in the world and more than the majority.

As we learned in the previous chapter, cheating and cheat detection is like a cat-and-mouse game, with one trying to outwit the other. As related by Abagnale, two of the most important steps a con artist should take are observation and research. Applying these himself, he was able to stay ahead of the game by finding cracks in the system—in people and institutions—and devising new schemes that no one else had tried. In other words, creative use of the Second Law was a key to his cheating success.

A Ponzi scheme is a good illustration of how sophisticated schemes for cheating can get past our guard. First and foremost, people can be seduced by opportunities to make quick money. Ponzi schemes are crafted to pander to this common bias. Yet, how they work is sufficiently complex and obscure enough to fool most of us. That's why most victims of Bernie Madoff's $16 billion Ponzi scheme were wealthy people, many of them normally savvy in business and financial affairs. The deceptive tricks that undergirded the scam were beyond the victims' comprehension from the outset.

The housing bubble in the 2000s was an example that more of us can relate to in the context of our own lives. Normally, real estate investors tend to be conservative, seeking stable returns with low risks. However, lured by years of steadily rising housing prices, many began to let down their guard and jumped on the speculative bandwagon with the false belief—a serious cognitive bias—that housing prices could only go up.

Such a bias was maximally exploited, as bankers fueled the real estate bubble and profited from it by lowering lending standards to increase loan volume. Investment firms launched new and complex financial products such as CDOs (Collateralized Debt Obligations, securities backed by assets such as real estate) to allow mortgages to be traded like stocks.[33] Credit rating agencies joined the parade by propping up the quality of the underlying loans in these products.

Scams were aplenty during the buildup of the bubble. Self-proclaimed investment gurus popped up in droves, offering courses promising "secrets" of making money in real estate. New and complex investment products were marketed to pension fund managers and institutional investors who knew little about the products' risks, which were hidden behind their complexity. The fact that Lehman Brothers—one of the most respected investment firms on Wall Street—could go belly-up illustrates that even seasoned investment professionals didn't recognize the risk of putting money in these "toxic assets." This shows that financial products can be so intricately designed as to be beyond the comprehension of most. Yet they are often purported to be golden profit opportunities—before their true color of fool's gold is revealed.

Meanwhile, violations of financial regulations and laws, overt or covert, were rampant. Banks sold loans to unqualified customers, whereas investment firms repackaged and resold those loans as risky financial products to uninformed investors. Few of the financiers, however, suffered the consequences of their irresponsible, unethical, or unlawful deals. In the United States, amazingly, none of the Wall Street CEOs were convicted. This shows that cheating, at least by established institutions, is often "safe" to commit. Those responsible can easily go undetected or get away unpunished. How can you prevent cheating when its cost to the cheater is trivial or nonexistent?

Today, some of the schemes Abagnale concocted may be out of fashion, but his creative use of the two laws of cheating continues to work, especially in the digital world. Despite the vast diversity in Internet scams, one common feature is their novelty, which can enable these scams to slip past our guard until it's too late.

As more and more business transactions are done online, the economic stakes are getting higher in the digital space. Meanwhile, the vast and far-reaching capacity of the Internet provides unlimited opportunities for inventing new scams. Yet our own cognitive ability to detect them has improved little since the Stone Age. Using our obsolete mental tools to fight fast-evolving Internet scams is like sending medieval cavaliers to charge at modern enemies armed with tanks and missiles. Clearly, we're on the short end of the stick in defending ourselves against a new generation of scammers and swindlers who are invisible, innovative, and skilled in cutting-edge technologies. (We do, however, have some ideas for fighting back, but let's save this topic for the last chapter.)

𝕬

How do cheaters choose their prey? Abagnale selected his victims carefully for his check frauds. He preferred to deal with naïve-looking people such as young female bank tellers who were easily distracted by his handsome physique, flirtatious demeanor, and apparently respectable profession. In his own words, "It's not how good a check looks but how good the person behind the check looks that influences tellers and cashiers."

Selection of victims can be more systematic. Consider this. Many of us have received emails from some Nigerian princes, pledging to send millions of dollars for a joint business venture. If you fall for the offer, you'll be asked to pay a few hundred dollars as "the processing fee."

The scam is so transparent that you may wonder why the putative princes would even bother to send the message—again, again, and again. The scammers are by no means stupid. On the contrary, the messages that seem so plainly fake to most of us are intentionally crafted to screen for people who have the right mental loopholes. The logic: if you can't see the obvious problem in the message, your judgment is sufficiently impaired to make you a potential victim. In other words, the scammers prey on people who lack the judgment—those who have mental loopholes that make them miss the obvious. It's not difficult to guess that most of their victims are the elderly with declining cognitive abilities.

The Nigerian prince scam is an example of what I call *individual cheating*, directly targeting fellow humans—spouses, friends, relatives, coworkers, acquaintances, business partners, total strangers—for resources such as money, sex, and social status. Individual cheating surprises nobody because a wide range of animals cheat in just this way, as we all resort to the two laws of cheating that are quite familiar to us now.

What sets humans apart from other animals—including our close primate relatives such as chimps and bonobos—is what I call *institutional cheating*. This is cheating on rules and systems—taxes, voting processes, educational tests, or business opportunities. In this kind of cheating, the victims are not individual humans, but impersonal organizations— including firms, schools, NGOs, and governments. This was what Abagnale's frauds were mainly aimed at. "My targets," he confessed, "had always been corporate targets—banks, airlines, hotels, motels or other establishments protected by insurance."

The prevalence of our social, economic, and cultural institutions opens a whole new world of opportunities to cheat. The victims are ultimately still people, but their individual identities are often unknown or difficult to define, so they are far less likely to provoke our sympathy. For example, if Bank of America loses a million dollars, you may just shrug and say, "That's a shame!" But your feelings would be totally different if $10,000 were stolen from John Doe or Jane Smith, who in your mind is a kind person with a ready smile. For that reason, cheating on institutions may not carry the same moral status as cheating on individuals, as Abagnale experienced. At times, cheaters may even feel justified, like Robin Hood, if the organization they attack has a less-than-stellar reputation. These are additional reasons why institutional cheating is widespread.

A well-known example of institutional cheating—that is, cheating directly on rules but indirectly on other competitors—is doping in competitive sports; among the best known doping cheaters are Lance Armstrong (cycling), Maria Sharapova (tennis), Diego Maradona (soccer), and Marion Jones (track and field). Doping breaches the spirit of fair play, but high stakes—fame and commercial profit—can still motivate some to cheat. For example, in 2011 and 2013, as many as 18% and 15% of athletes, respectively, used banned drugs to enhance their performances in the World Athletics World Championships.[34] The situations are similar for the Olympic Games and many professional sports. As dopers often employ the most advanced technologies, their scams may go undetected for years. By the time the scams are revealed, the athletes will have already made their fortunes, and scandals will mostly become irrelevant. That's why doping in sports is so hard to stop.

Free riding is a far more common form of cheating than scams, particularly in large organizations. Institutional free riding is easy to get away with when individual contributions are mingled with team efforts. For example, one specific teacher can't be held responsible for the educational quality of an entire school. In many situations, free riding is tolerated when its impact is trivial, for it's not worth the effort to police. Even though free riding may be obvious and people who contribute their effort faithfully may hold a grudge, the institution may not be able to do much about it. How many of us haven't experienced just such a situation at work?[35]

Nevertheless, institutional cheating in general and free riding in particular can destroy team spirit and, in some circumstances, even endanger other members of the team. Consider the case of Robert "Bowe" Bergdahl. On June 30, 2009, he deserted his duty and walked away from his post in Afghanistan. He was captured and held by the Taliban for the next five years before he was swapped for five Taliban captives detained at Guantanamo Bay in a deal brokered by the Obama administration. Many Americans felt outraged at the "bad deal," as they believed Bergdahl, a deserter, was not "worth" it.[36]

Deserters are clearly cheaters, if not outright traitors. The fighting ability of a military unit relies on the absolute loyalty of every member. A rogue soldier like Bergdahl can put his team members in grave danger and weaken the fighting capacity of the unit. The stakes are so high that cheating must be absolutely suppressed. Thus, punishment for military deserters is often extremely harsh, including summary execution in some cases.

Institutional cheating can also be committed by organizations as a whole. A case in point was the cheating on emission tests by the auto giant Volkswagen. In 2015, several of its models including the Beetle, Jetta, Golf, Passat, and Audi were found to be equipped with a sophisticated cheating apparatus known as "the defeat device," which gave lower readings of carbon emissions. After they were caught, the ensuing scandal cost the company $7.4 billion in recalling 350,000 "faulty" vehicles, and $80 billion in the company's market value in Europe alone. The loss might be considerably larger because Volkswagen is listed in several markets around the world. Its business was badly impacted, forcing the company to cut back its worldwide operations by selling several factories and other assets to pay expenses.

Why would a gigantic corporation like Volkswagen cheat? Couldn't they recognize that there was so little to gain yet so much to lose? The answer is that organizations are run by individual people. And in most for-profit organizations, many are rewarded for short-term performance, usually measured by annual or even quarterly profit or loss. In addition, the social milieu of large organizations can facilitate free riding, allowing individuals to diffuse their personal responsibilities if anything goes wrong.

Put yourself in a similar situation. You're evaluated annually for your performance, which determines your salary, bonus, and promotion. You discover that if something goes wrong, you can easily find a scapegoat. You may tell yourself at first, "I'm going to hold my moral ground as an honest person." But before long, you discover your coworker Joe, who has an unsavory reputation of using tricks and taking credit from others, has just received a raise, apparently due to his "superior performance." The system is against you. So, this is a no-brainer: when the downside of cheating is minimal, cheating can prevail.

What if Joe becomes John Stumpf, the former CEO of Wells Fargo? To falsely enhance apparent performance, Wells Fargo created millions of fraudulent accounts. This illegal practice was at least sanctioned, if not directly ordered, by Stumpf himself. After the scandal was exposed in 2016, the company had to pay $2.7 billion in fines and lawsuits.[37] Stumpf, however, was only forced to step down.

Stumpf was unlucky because the scam was revealed while he was still in charge. More often than not, such corporate malfeasance may go undetected for years, like doping in sports. By the time scams are discovered, company executives have already retired or moved on. Why would they be afraid of getting caught? That's why tobacco CEOs were so ready to lie when they were summoned for a congressional hearing in the 1990s. When asked whether nicotine is addictive, they all gave the same answer: "I don't believe nicotine is addictive," despite the fact that they were all fully aware that their own internal research contradicted their claims. This example makes clear another aspect of institutional cheating: if a fraudulent practice is widespread, it becomes a norm that carries little or no risk. That was exactly what happened when the real estate bubble burst in 2008. What if the heads of big banks had to personally pay for the colossal losses or go to jail? Would they still have been so ready to cheat?

Institutional free riding can even bankrupt an entire nation, as illustrated by Greece's sovereign debt crisis in 2011. Greece was relatively poor compared to other European nations. Yet, since the 1990s, the government boosted the country's living standard by borrowing excessively, allowing more and more Greeks to buy nice houses, drive luxury cars, and take exotic vacations. Furthermore, the government piled up mountains of debt to provide an exceedingly generous pension system, give comfortable benefits to the unemployed, and fund bonuses for civil servants. It even launched a "Tourism for All" program, handing out free money to low-income people to take vacations.[38] Reckless spending by borrowing made the politicians popular among voters, but it made the entire Greek economy look like (in fact, it was) a gigantic Ponzi scheme for unsustainable extravagance.[39] Should we be surprised by a bad ending for the Greek economy after years of free riding gone wild?

🔏

Institutional free riding is far more widespread than many might think. It's common whenever there are organizational loopholes. Such loopholes can exist in a loose association or a well-organized union, a private company or a public school, a small township or a large nation.

Let's use American colleges as an example. As recently as the 1990s, the administrative structure of most higher education institutions was quite simple, typically with a president, one or two vice presidents, and one provost, at the top level. Today, "universities are filled with armies of functionaries—the vice presidents, associate vice presidents, assistant vice presidents, provosts, associate provosts, vice provosts, assistant provosts . . . each commanding staffers and assistants," laments political scientist Benjamin Ginsberg in his book *The Fall of the Faculty*. Moreover, most of these professional administrators have little interest or experience in teaching or research but see "management as an end in and of itself."[40]

In the two decades before 2011, Ginsberg writes, the number of faculty members and students increased by about 50%, but the full-time administrators proliferated by 85%. "Between 1997 and 2007, the ratio of administrators for every one hundred students increased by about 30 percent at private colleges. . . . During the same period, however, some universities—Yeshiva and Wake Forest, in particular—experienced more than 300 percent growth in the size of their managerial and support staffs."[41]

David Graeber at the London School of Economics shared this concern, writing scathingly on May 6, 2018, in *The Chronicle of Higher Education*, "As managerialism embeds itself, you get entire cadres of academic staff whose job it is just to keep the managerialist plates spinning—strategies, performance targets, audits, reviews, appraisals, renewed strategies, etc, etc.—which happen in an almost wholly and entirely disconnected fashion from the real life blood of universities— teaching and education." Why so?

The short answer is bureaucracy, which has become so widespread that it's often mocked as a system of parasites, with bureaucrats forming recursive relationships, described in a rhyme composed by the nineteenth-century mathematician Augustus de Morgan:

> Great fleas have little fleas upon their backs to bite 'em,
> And little fleas have lesser fleas, and so ad infinitum.
> And the great fleas themselves, in turn, have greater fleas to go on;
> While these again have greater still, and greater still, and so on.

Because bureaucracy can become a major societal venue for institutional cheating by free riding, it often carries an off-putting connotation of inefficiency, redundancy, officious rules, and unnecessary procedures. Even so, bureaucracy is not intrinsically bad. On the contrary, it's essential for running any organization, especially a large one. This was how Max Weber, the German sociologist of the late nineteenth and early twentieth century, saw it.

Weber believed that bureaucracy, if it works well, can preserve order, promote efficiency, eradicate favoritism, and reduce transaction costs in the economy. That's why modern bureaucracy has emerged in both public and private sectors, including government administration, military units, churches, political parties, public and private corporations, colleges and universities, professional associations, and NGOs.[42] Weber embraced the rise of bureaucracy as a progressive milestone in Western civilization.

Weber's rosy view of bureaucracy was not shared by his contemporary, the novelist Franz Kafka, who once served as a low-level officer at the Workers' Accident Insurance Board of Bohemia. Kafka recounted his experience in several of his fictions, most vividly in *The Trial*, *The Kastle*, and *In the Penal Colony*. Those three literary works were exposés of bureaucratic inefficiency, incompetence, cruelty, and abuse. Bureaucracy, for Kafka, was an excuse for officials to free ride, taking advantage of the system to advance their own interests. He eventually lost his faith in all forms of bureaucracy, writing in deep disgust: "Every revolution evaporates and leaves behind only the slime of a new bureaucracy."

Between these two opposed perspectives on bureaucracy, who was right, Weber or Kafka? The answer is complicated. Weber was a theorist of modern government, which, he believed, could run like a well-oiled machine. He envisioned a working bureaucracy to have the following five basic elements, as summarized by sociologist Randy Hodson and his colleagues:

a) a *hierarchy* with a clear chain of command,
b) exhaustive *written rules* to govern all regular operations,
c) specialized *departments* for technical efficiency,
d) formal *training* for bureaucrats in their areas of expertise, and
e) well-defined duties that require the *full capacity* of the official.[43]

For Weber, a bureaucracy *ought to* show the following characteristics: "Precision, speed, unambiguity, knowledge of the files, continuity, discretion, unity, strict subordination, reduction of friction and of material and personal costs."[44] In the real world, however, bureaucracies consist of, and are run by, people with self-interest, not Weber's cold, emotionless machines. For this reason, each of the Weberian elements can be bent, abused, or violated outright, making bureaucracies hosts for institutional free riding. Here is how.

We turn to the hierarchy element first. Bureaucratic hierarchies are, in theory, set up for streamlining workflow by clearly defining responsibility and accountability for each member. But individual bureaucrats tend to be oblivious of this big-picture view, focusing, instead, on how they can survive in the system and realize their personal career ambitions. For instance, officials may put a greater priority on pleasing higher-ups than on serving the public. That's why bureaucrats may come off as arrogant and unsympathetic.[45]

Unit leaders may also use their positions to expand personal power and prestige by obtaining bigger budgets and hiring more people.[46] Therefore, bureaucracies tend to grow at the cost of efficiency,[47] as keenly observed by British Naval historian Cyril Northcote Parkinson. In a 1955 essay published in *The Economist*, he wrote that "work expands so as to fill the time available for its completion." This has been dubbed, somewhat tongue-in-cheek, Parkinson's Law. The net effect is that more people end up doing the same amount of work. In his ensuing book published in 1957, Parkinson calculated that a bureaucracy in the British Civil Service grew in size at a consistent rate of 5–7%, regardless of the amount of work it undertook. Two ingrained ideas that drove its growth were: "an official wants to multiply subordinates, not rivals" and "officials make work for each other."[48]

As more and more people are added into a working unit, it will eventually reach a point where there are more people than the amount of work to do. What happens next is that the bureaucrats and office workers pretend to be busy so they don't seem to be slacking off. But if you think free riding ends here, you're wrong. Bureaucrats need a sense of self-importance, too. They may derive it by keeping others busy as well. "In most universities nowadays," David Graeber writes, "academic staff find themselves spending less and less time studying, teaching, and writing about things, and more and more time measuring, assessing, discussing, and quantifying the way in which they study, teach, and write about things." He estimates that "at the very minimum, 90% of the role [for him as Head of Department] is bullshit." Does this feel familiar?

Hierarchy also generates incompetence. Since rank begets money, power, privilege, and prestige, bureaucrats see promotion as a gauge of personal success, whether or not they are suited for higher positions and capable of performing their duties. Thus, bureaucrats continue to climb the rungs of the organizational ladder until they reach their maximum level of incompetence. This is known as the Peter Principle, discovered by psychologist Laurence Peter.[49]

Now, let's turn to the second element of a bureaucracy: written rules. For Weber, written rules and documentation should be used to preserve regularity and transparency in an organization. The reality, however, often falls far from this ideal, for two reasons. First, policies can be poorly administered because of their ambiguity or obscurity.[50] This allows knowledgeable officers—especially those at higher levels—to manipulate information in their personal favor. Second, as new policies are progressively added to the regulatory structure, complexities also rise over time,[51] leading to new loopholes. This in turn prompts the implementation of even more new policies in a misguided attempt to close the loopholes. As the cycle continues, a bureaucracy may create more and more opportunities for free riding even though the intent is to suppress them.

Let's now examine the third element: the division of departments, purported to increase efficiency and quality of service. In reality, however, each unit is prone to expand its turf to gain more power and control, as we've seen. This can lead departments to encroach into one another's territories, resulting in more and more overlap in duties and personnel. Take the US intelligence community as an example. It's made up of 17 organizations under seven federal departments with a tremendous amount of redundancy in function and staffing.[52] Anyone could likely find opportunities to free ride in a complex and tangled system such as this.

Finally, the fourth and fifth elements of a bureaucracy—professionalism—may also be corruptible. Bureaucrats are supposed to serve to their full capacity. Their incomes are mainly, if not fully, based on performing their professional roles. But the power they gain from assuming their positions is an invisible resource that encourages officials to establish personal relationships or even offer personal services, leading to blurring of the lines between public and private life for bureaucrats.[53] One example is the "cozy" relationship between Wall Street and market regulators in the SEC (Securities and Exchange Commission). Research shows that the SEC tends to investigate cases of backdating stock options and lower-stake cases. This might result in lower probabilities for individuals or companies to be punished.[54]

Also, bureaucrats are expected to be experts in their respective fields and exhibit professionalism in their posts, but their capability and performance are often beyond the knowledge of their supervisors.[55] The quality of a bureau's work is often hard to judge, much less to monitor and censure for poor performance.[56] Because there is little consequence if they don't do their jobs well, bureaucrats have few incentives to work diligently, except for pleasing those above them through whatever means, which includes cheating.

Even though leaders push for certain policies, bureaucrats can still come up with shortcuts to artificially enhance their performance just so they can look good.[57] In Texas, for example, when a school district's performance was measured by standard test scores, administrators and teachers in some schools cheated by preventing poor-performing students from taking those tests. On average, 9.2% of the students were excluded, but in some districts, the percentage of test exemptions could reach 35%.[58] This does not even include a more subtle form of cheating: teaching students to perform well on tests at the expense of other subjects and activities.

Moreover, requiring bureaucrats to be experts in their jobs can be jeopardized by political loopholes. For instance, an American president may pay back political supporters by giving a powerful position such as Secretary of Education to a person who has no experience in the public education system, or Secretary of Energy to a person who advocates the elimination of the very agency he leads.[59] When many key positions are occupied by unqualified and incompetent personnel, the government is flirting with grave danger. Author Michael Lewis calls this pervasive problem the "Fifth Risk" in his 2019 book of the same name.

If top-level officials get jobs through patronage and favoritism, lower-level positions may also be filled in the same way, often going to relatives, friends, or those sharing political views with recruiters. (Today, many administrative jobs are created explicitly to "serve the pleasure" of the higher-ups. If this is not free riding, what is it?)

Although Weber and Kafka may appear at odds with each other regarding the workings of bureaucracies, their views are actually two sides of the same coin. Weber served as a director of nine army hospitals in Heidelberg during World War I, which gave him a top-down, optimistic outlook about government administration. Kafka was a low-ranking, marginalized clerk. He saw bureaucracy from a bottom-up, pessimistic perspective: inefficiency, arrogance, and corruption. What Kafka took as reality was what Weber believed as the worst-case scenario: human lives could be trapped in the "Iron Cage" of bureaucracy, straitjacketing personal freedom.[60] So, if Weber was a grand designer and dreamer of bureaucratic efficiency, Kafka was a physician who saw its pathological flaws.

Bureaucracy reminds me of my visit to the massive hydroelectric project at Grand Coulee Dam in Washington in the summer of 2019. During a tour to the generator room, our guide pointed at a gigantic machine and explained a small problem during its installation. "It could be fixed by one engineer," he said, "but it took six directors and many committee meetings." When I chuckled, he added, "This is how government works."

How to cure such bureaucratic ills? If you pin your faith on privatization, you're bound to be disappointed because bureaucratic inefficiency is a systemic problem, mostly independent of who runs it. Take, for example, the growth of administrative staff in higher education in recent decades. Private four-year colleges have far outpaced their public counterparts in this regard (fig. 6.2).

The same can be said for the bloated administrative offices in the American health care system, which is dominated by private companies. When you visit any clinic or hospital today, you must first deal with specialists who handle insurance issues. According to a 2010 study by the National Academy of Medicine, the portion of expenses that went to billing and insurance-related costs were twice as much as necessary. A 2017 estimate put the total cost of health care administration at $1.1 trillion—45.6% more in relative percentage than in France, which came in second for the cost of health care administration. On a per capita basis, administrative costs were $1,059 per year for Americans, compared with $307 for Canadians.[61] For the record, health care in France and Canada is universal, and Canada has a single-payer system.

Where does all that money go? According to David Graeber, to free riders, or what he calls people with "Bullshit Jobs," in his recent book with that title. He estimates office workers spend only half their work hours being productive. The other half is sucked into pointless tasks such as emailing, meetings, and meaningless administrative work. In European countries, 37–40% of people believe their jobs contribute nothing to society. They include such employment sectors as lobbying,

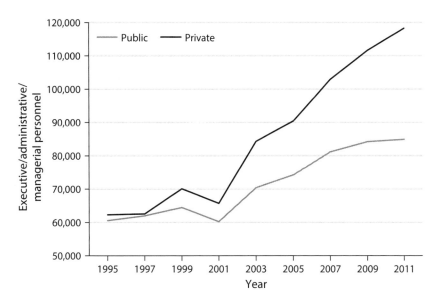

FIGURE 6.2. Growth in executive/administrative/managerial personnel in private and public colleges from 1995 to 2011 (data from NCES Digest 2018).

telemarketing, corporate law, and financial and management consulting. Graeber believes there would be no downside to society if half of all these jobs were eliminated.

Economic studies indeed demonstrate that many professionals are free riders, taking more from society than they contribute. Here's a breakdown for the dollar value society gets back for every dollar paid out in salary to professionals (negative indicates net loss): $9 for medical researchers, $1 for schoolteachers, $0.2 for engineers, 0 for consultants and IT professionals, −$0.2 for lawyers, −$0.3 for advertisers and marketing professionals, −$0.8 for managers, and −$1.5 for the financial sector.[62] You can sense how big a problem free riding is for society!

What is also clear here is that privatization is not a solution for bureaucratic free riding. On the contrary, it may make the situation worse. The reason is no mystery. When profit reigns supreme, anything that can increase an organization's bottom line may be tolerated or even encouraged. A firm may be pressured by its stakeholders to try to save money by externalizing costs including pollution, health, and accidents during production.[63] To get things done, owners and managers may become obnoxious bosses, resorting to bullying tactics such as profanity, threats, selective enforcement of rules, or firing people for personal problems or temporary under-performance. This can lead to a great deal of anxiety and fear in the workplace.[64] Unlike public sectors that heed rules in labor protection, gender equality, and Affirmative Action, private firms are more likely to ignore these welfare and fairness considerations. Bureaucracies in today's private organizations, as reflected in their corporate cultures, tend to be more Kafkaesque than Weberian.[65]

How then can we deal with bureaucratic inefficiency? To answer this question, we have to find where the problem lies. For this, let's zero in on the US federal government.

𤩖

In his first inaugural speech on January 20, 1981, Ronald Reagan echoed the Kafkaesque populist view and roused his followers by making a sweeping statement: "In this present crisis, government is not the solution to our problem; government is the problem." This speech is often taken as an indictment of American federal bureaucracy for its excessive reliance on regulation, or "Big Government," a derogative term preferred by some conservatives. Does such a Kafkaesque view regarding American government have merit?

There is no doubt American government is large. The federal government alone consists of about 2,000 agencies with 2.79 million civil servants. In 2011, a DC-based organization called Citizens Against Government Waste aired a one-minute infomercial on TV, showing a charismatic Chinese leader speaking to a crowd while laughing at the US being brought down by big government and wasteful spending.[66] As provocative as it was, the infomercial contained a fatal error: the Chinese government is the largest in the world. It employs 7.17 million civil servants plus 31.75 million more on the government's tab. While the Chinese population is three times larger than the population of the US, its government is 13 times larger. I called the office of Citizens Against

Government Waste to point out the problem, and before long, the infomercial disappeared from TV.

Not only is the US government comparatively small, but it has not grown in relative terms since World War II. The federal government accounted for more than 5% of total employment in the 1950s. But today, the number has dipped below 2%, while the population has doubled and the GDP has grown by more than seven times since then.

So, if size is not to blame, where is the problem? The answer: structure. When FDR created the modern American government in the 1930s, it was effective and efficient, responding well to the needs of winning World War II. Since that time, the government has been continually downsized and defunded, yet its structure has grown increasingly complex, laden with inflexible rules and mandates. For example, Kennedy's cabinet departments had only 17 layers in their administrative structure. When Trump took over, he faced 71 layers in the bureaucratic hierarchy.[67] Even China, with an administration many times larger, does not have such a complicated system. Facing this convoluted bureaucracy, Trump did what many conservatives wanted him to do: slashed several agencies, including, unfortunately, the pandemic response office in 2018. What had gone wrong with Trump's approach?

Alexander Hamilton once said, "A government ill executed, whatever it may be in theory, must be, in practice, a bad government." But in normal times, it's hard to see whether a government is good until there's a crisis that provides a rigorous test of government efficiency. However, even before the onset of COVID-19, the US government failed several such tests. These included Hurricane Katrina in 2005, which killed over 1,200, and Hurricane Maria in 2017, which killed more than 3,000.

Unfortunately, these crises failed to draw enough attention to the bureaucratic troubles in the US government until the COVID-19 pandemic ravaged the nation, exposing its problems in leadership, preparation, and response. Even basic medical equipment such as ventilators and personal protective gear were in seriously short supply, though China had already shown how vital they were at least six weeks earlier. Most absurdly, the government was even unable to make Americans wear face masks in public places to slow down the spreading of the virus.

The main problem, according to CNN commentator Fareed Zakaria, is that "federal agencies are understaffed but overburdened with mountains of regulations and politicized mandates and rules, giving officials little power and discretion." Apparently, "both parties have contributed to the problem, making the federal government a caricature of bureaucratic inefficiency."

When a bureaucracy is so inefficient that it can't perform its function well, the system is broken, becoming de facto an ideal host for free riders. That is, when free riding is institutionalized, people in the system may have no choice but to go with the cheating tide, even though many may want to contribute to society's greater good. Thus, it is not necessarily that our government is too big, nor that too many people want to ride for free. The system is simply too encumbered to effectively do its job. If we can understand this key problem, the solution becomes obvious: simplify the structure by flattening the hierarchy and reducing its layers of bureaucratic complexity; don't simply slash its size without discretion as Trump did.

In this chapter, we've examined human cheating prompted by three questions: What do cheaters cheat for? How do their schemes work? And whom do they prey on? We've found that human cheating is both universal and unique. In terms of universality, humans are driven by the same instincts and use the same rules to lie (First Law) and deceive (Second Law) as other animals do. In terms of uniqueness, human cheating keeps pace with changes in our culture, for which it goes far beyond the reach of any other biological species in diversity, complexity, and ingenuity. Moreover, humans can cheat individually, a feature shared with all other animals, and institutionally, which is unique to humans.

To examine institutional cheating prevalent in virtually all organizations, we zeroed in on government bureaucracy and took a stab at understanding how it can become a haven for free riding. Unfortunately, how to overcome bureaucratic inefficiency is a topic too large to fit in just a few pages here. Therefore, we leave the question to social scientists, especially scholars in government public administration, while moving on to the next big issue in cheating: self-deception.

CHAPTER 7

Liars Who Lie to Themselves

When you know a thing, to hold that you know it; and when you do
not know a thing, to allow that you do not know it—this is knowledge.

—CONFUCIUS

The meaning of the phrase "Know Thyself," inscribed in the Temple of
Apollo at Delphi, has inspired a spirited debate among classics scholars.[1]
But the true import of this ancient dictum had not been broadly appreciated in America until the founding of the Minnesota town of Lake Wobegon in 1974. Despite a small population of 900 people, the town was special
because "the women were strong and the men good-looking and the
children all above average," as told by Garrison Keillor.

Keillor confessed in 2014 that he invented the town and its inhabitants
for the "News from Lake Wobegon" segment of his radio program, *A
Prairie Home Companion*, aired on hundreds of public radio stations. For
42 years, from 1974 to 2016, millions laughed heartily at the stories from
Lake Wobegon, where people often think they are better than they
really are and do foolish things beyond their capabilities.

Although a comedian, Keillor wasn't really joking. People in most
towns aren't so different from those in Lake Wobegon. The above-
average effect, known as illusory superiority in psychology, can be

found in all aspects of our lives. That's why the show was popular for decades. In fact, the fictional stories Keillor told felt so authentic that many in his audience believed Lake Wobegon was real. Apparently, many of us have little appreciation for the limits of our own knowledge and abilities. Instead, we're inclined to overestimate both. That is, we cheat on ourselves.

The prevalence of self-deception is truly staggering. In regard to our personal health, for example, most people believe they live a healthier lifestyle and have a longer lifespan than their peers.[2] Over 90% of people believe they are better-than-average drivers.[3] In social skills, 70% of high school students consider themselves above average in leadership, and 25% blatantly put themselves in the top 1%.[4] Likewise, most people exaggerate their popularity and inflate the number of their friends.[5] In academic and job performance, 87% of students rate themselves better than their average peers, and over 90% of faculty members place themselves in the top half in teaching ability.[6] The same can be said for lawyers who think they can win a case or for stock traders who consider themselves to be the best in the business.[7]

Under the spell of self-deception, people overstate their incomes, attractiveness, happiness, technical skills, biological endowments, and moral character. They often unwittingly brag, selectively presenting only the good side of themselves in school, at work, and online. How many of your Facebook friends, for instance, post pictures or video clips about the downside of their lives—such as being demoted, having financial troubles, or getting dumped in a relationship?

Self-deception often compels us to attribute successes to our own effort, skills, or intelligence, but we excuse our failures as due to external causes or problems on the part of others. "Mistakes are made," we may claim when things don't go well for us, instead of stating the simple fact, "we're wrong," or "we've flunked." Even when there is nobody to lay blame on, we may still look for a scapegoat: we split ourselves into past and present personalities, then claim that our past self didn't do so well, but our present selves are doing much better.[8] We are new people now.[9]

The same narcissistic tendency makes us like the images we see in a mirror more than those captured in photographs because mirror reflections are mostly appreciated by ourselves only, while photos can be viewed by others.[10] For the same reason, we're quicker to pick out our own image when it's artificially enhanced to be more attractive than when it's not.[11] Apparently, most of us are more or less living our own lie.

Self-deception is so common in America that Mitt Romney famously used the political slogan "Join the 1% [of the richest Americans]" to appeal to voters during his presidential bid in 2012. (Romney was not an outlier, of course. Most campaign slogans—from Obama's "Yes We Can" to Trump's "Make America Great Again"—serve the same function of boosting voters' morale and self-confidence.) Clearly, many people are unable to recognize, much less admit, the ceiling of their own capabilities. (We will see later that women are more likely than men to play down their own abilities.) Otherwise, how could most be above average—to a degree that the very term "average" has lost its statistical meaning?

In the rest of this chapter, we will try to answer why we humans (and maybe other animals[12]) cheat on ourselves, how prevalent and diverse self-deception is in society, what some positive (such as high self-esteem and the placebo effect in healing) and negative ramifications (such as confirmation bias and overconfidence) are, and how we can overcome overconfidence.

๙

Psychologists have been engaged in a major effort to understand self-deception since the 1990s. One of the notable studies was done by Justin Kruger and David Dunning at Cornell University. The duo recruited 65 regular human "guinea pigs"—psychology undergrads—and asked them to estimate their abilities in answering questions about humor, grammar, and logic before they knew their real scores. As it turned out, participants who did poorly rated themselves far higher than their actual performance. This cognitive distortion was worst for those in the bottom quarter, who overrated themselves by more than 45% to be near the 60th percentile (fig. 7.1).

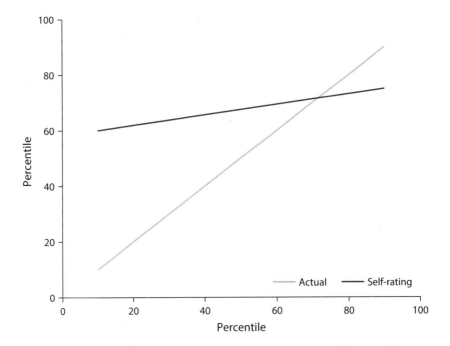

FIGURE 7.1. Performance based on self-rating and actual score (modified from Dunning and Kruger's original research).

Kruger and Dunning published the research in a 1999 paper, "Unskilled and unaware of it: How difficulties in recognizing one's own incompetence lead to inflated self-assessments."[13] This ignorance of one's own ignorance came to be known as the Dunning-Kruger effect, or, more pedantically, meta-ignorance.[14] It's probably easier to remember it in crude lingo: idiots don't know they are idiots.[15] Self-deception is the reason there is often a significant gap between our own assessment of ourselves and that by our peers. It is, in Dunning's words, "a double burden," preventing many of us from knowing why we make mistakes and commit errors, especially for low performers.[16] But if deceiving ourselves won't do us good, why do we still do it?

Interestingly, the first person who put a serious effort into answering this question was not a psychologist but an evolutionary biologist: Robert Trivers. Trivers noticed the dilemma as early as the 1980s.

Self-deception has an obvious cost. At times, it may lead to family disputes, personal romantic disasters, airplane crashes, or even going to war—the second Gulf War, as a notable example. For self-deception to thrive in evolution, Trivers thought, there has to be some biological benefit to offset its cost. How might self-deception have increased a person's chances of survival and reproduction in our past?

As we know, when people lie, they often betray their dishonesty through their own behavior, especially by their facial expression. Lying is difficult when we're conscious of making up a false story. This is because our brains are poor at multitasking. In my own experience, I have no difficulties speaking English or Chinese, but only one language at a time. If I quickly switch back and forth between the two languages, my fluency in both languages is greatly impeded. This happened when I served as an interpreter for a 2004 Sino-American joint symposium in Hefei, China. I often stuttered as my mind went momentarily blank while trying to come up with exact words or expressions in the other language.[17]

The same can be said for intentional lying. When you lie while fully aware of what you are doing, you force your brain to play Dr. Jekyll and Mr. Hyde simultaneously. This task is far harder than finding parallels between two languages: you must deal with contradictions between reality and the words coming out of your mouth. To quash the truth, your brain has to handle an extra burden, known as cognitive load, which can make you overcontrol what you normally would do. As a result, you may tense up and behave too rigidly. You talk with a higher pitch; you make longer pauses in your speech; and you fidget, gesture, and blink your eyes less than normal. These and other features, such as unusual facial expressions and behaviors, can easily betray you.

Cognitive load is best illustrated by lie detection in criminal investigations through what is known as forensic linguistic analysis. Criminals—even those who have carefully rehearsed their lies—often give themselves away during interrogation. They can be extremely nervous because they are fully aware of their crimes, so they often use different terms than they usually would. Their speech patterns deviate from the normal as well, with fewer qualifiers and more negative words. When pressured, they often blurt out what they are trying hard to hide. Here is an example.

On October 7, 2017, a Georgia couple, Christopher McNabb and Cortney Bell, reported that their two-week-old daughter Caliyah was taken from their home. The couple pleaded to the community for help. The following is part of the conversation between the couple when they were left alone in a police interrogation room:

> "I loved—I love her. She was my baby," says McNabb to Bell.
> "Calm down. Why did you just do that?"
> "Do what?"
> "Why did you just do that?"
> "Why did I just do what?"
> "You just said 'loved.'"
> "I don't know, Cortney. Where is she? You think I had something to do with this shit?"
> "In my heart, no. I just hope you haven't. My heart tells me no."[18]

The couple didn't realize that two words in their conversation tipped off the police. Do you see them? The use of "loved" and "was" in the past tense when the conversation started signaled that they knew their baby daughter was already dead before the interview took place. Although McNabb quickly corrected his Freudian slip, Bell was immediately alarmed by its potential consequence.

The police searched the woodland near the couple's mobile home and found Caliyah's body wrapped in a drawstring Nike bag. Numerous skull fractures indicated she was brutally beaten to death. As it turned out, the couple had a history of violence and had killed their baby while high on meth. They were indicted for multiple counts of murder and assault, convicted on May 14, 2019, and sentenced to life in prison for McNabb and a 30-year jail term for Bell.[19]

You don't need to put yourself in the shoes of a criminal to understand the burden of cognitive load. Conflicts that result from internal contradictions are easily identifiable by people in our everyday lives, especially those who know us well. How many of us, for instance, have tried to smile even when we're not pleased? Such a forced facial expression is known as a social smile in contrast to the honest Duchenne smile (fig. 7.2), named after the nineteenth-century French neurologist Guillaume Duchenne, who studied emotional expressions.

FIGURE 7.2. Stephen Colbert and Steve Carell. Which smile is a Duchenne smile, and which smile is fake? (copyright: Getty Images).

Is there a way to ease cognitive load so you can be a better liar? The answer is yes, and the recipe is to believe in your own lies. If you do so, your mind will no longer be burdened by the conflict between what's real and what's fabricated. Instead of betraying you, your expressions and behaviors become accomplices to your lies, simply by staying normal. This line of thinking led Trivers to hypothesize that "we lie to ourselves the better to lie to others."[20] In other words, self-deception evolved to "make deception more difficult to detect," Trivers writes. "Self-deception occurs when the conscious mind is kept in the dark. So, the key to defining self-deception is that true information is preferentially excluded from consciousness and, if held at all, is held in varying degrees of unconsciousness."[21]

Furthermore, even if your falsehood is revealed, it's easier to defend your innocence and sincerity if you yourself believe it. If you have no intention to lie, you won't lose your credibility. Thus, cheating on yourself, unlike cheating on others, has little social consequence.

Indeed, your mind functions more smoothly when you believe your own lies. Brain scans show that when people see themselves as more desirable than others (that is, they are deluded by self-deception), their medial prefrontal cortex experiences a higher level of activity,[22] whereas their orbitofrontal cortex and dorsal anterior cingulate cortex shut down. Apparently, the coordinated activities in these brain regions assume the role of "cognitive control."[23] It's unsurprising to note that these signature brain activities can go through the roof in a special group of people who hold a grandiose view of themselves all the time: narcissists.

✗

Back to the Dunning-Kruger effect: it's far more common and widespread than you might think. Setting aside students overrating their test performances, people generally overestimate their capacity in many aspects of life. This includes such things as reading comprehension, firearm safety information for hunters, research knowledge for lab technicians, medical diagnoses for physicians, technical skills for engineers, and competitive abilities for athletes. It may considerably change our perception on the quality of goods and services we receive when we realize the prevalence of self-deception in society at large.

Self-deception allows us to create our own subjective image to boost our self-esteem and self-confidence. For example, many women use makeup, perfume, plastic surgery, or breast augmentation to make themselves look younger and more attractive. Men, on the other hand, tend to resort to external enhancements—from gadgets they use to cars they drive—to make themselves appear wealthy and dominant. How much of our public image has *not* been altered to make it appear more positive?

Photo editing tools give us another powerful method to enhance our public image on the Internet. In this digital version of Lake Wobegon, men often appear unrealistically handsome, and women, drop-dead gorgeous. Many who see these unrealistically enhanced images fantasize about them and believe they are within reach.

Self-deception allows us to create our own subjective reality, more than simply physically or digitally altering our self-image. For example, many of us feel a meat product is healthier when it's labeled "95% fat-free" versus "5% fat." The same bottle of wine is judged as more desirable when it costs $90 instead of $10. Even the orbitofrontal cortex of the brain involved in choosing the wine indicates a higher level of enthusiasm for the one that's higher priced.[24]

We invent terminology and narratives to change our and others' worldviews. We can see this in the proliferation of euphemisms. "Anti-abortion" becomes "pro-life"; "global warming" becomes "climate change";[25] "civilian casualty" becomes "collateral damage"; "torture" becomes "enhanced interrogation"; "misfire" becomes "friendly fire"; "abduction" becomes "extraordinary rendition"; "genocide" becomes "final solution." And so on. The same feel-good makeover can be seen in public service campaigns. A road sign such as "Don't Litter!" becomes a fuzzy, warm feeling of "I♥NY" or a rough, tough warning of "Don't Mess with Texas."

Although it may promote self-esteem and self-confidence, self-deception can also enhance our pride and prejudice, leading to denial of facts and defiance of reality when they contradict our wishes and preferences. This is nicely illustrated in a simple study, where people were told to spit onto a paper strip and watch for the color change. The results show that those who believed that color change was good watched the strip 60% longer, hoping the change would indeed take place, than those who thought it was bad.[26]

As Trivers points out, our brains evolved to promote our own interests—to such a degree that our brains deceive us when deception best accomplishes that goal. As such, our memories can be created, re-created, edited, or manipulated to serve our own purposes. That's why our grandpas often rave about the "Good Old Days."[27] That's why we tend to remember far more details about our successes than failures. Memory, according to psychologists Carol Tavris and Elliot Aronson, can become "our personal, live-in, self-justifying historian," rewriting history "from the standpoint of the victor." Moreover,

> memories are distorted in a self-enhancing direction in all sorts of ways. Men and women alike remember having fewer sexual partners than they've actually had; they remember having far more sex with those partners than they actually had; and they remember using condoms more often than they actually did. People also remember voting in elections they didn't vote in; they remember voting for the winning candidate rather than the politician they did vote for; they

remember giving more to charity than they really did; they remember that their children walked and talked at an earlier age than they really did.[28]

Since a person can't be a passionate activist and a cool observer at the same time, memory's preordained mission in promoting our interests and protecting our feelings disqualifies it as a reliable storage device for information. Unlike a computer's hard drive, our memories can be quite distorted—to the degree that some claim vivid memories of out-of-body experiences, previous lives, going to heaven, or being kidnapped by aliens. Memories can make us feel like generous and responsible citizens by falsely recollecting our history of voting or donating to charities.[29] In the same way, when something bad happens, we may claim post hoc, "I said so," or "I warned you," when in reality, we never said any such thing. Memories are so easily manipulated that you can instill a false narrative in others' minds, which they will come to believe as truth. You can make people believe a fabricated story simply by recruiting one of their close relatives as an accomplice to confirm its authenticity.[30]

While these common phenomena may appear harmless, court testimonies based on false memories or confessions can be devastating. A recent review of cases shows significant problems in the reliability of witnesses' memories. This is particularly sticky in cases with children as eyewitnesses, histories of sexual abuse, and eyewitness identification.[31] Children are especially susceptible to the influence of suggestion, including reinforcement, repetitive questions, invitations to speculate, pressure from other witnesses, and introduction of new information.[32]

Many of us may employ false narratives to fool, comfort, or excuse ourselves. Trivers explains how this works:

> False historical narratives are lies we tell one another about our past. The usual goals are self-glorification and self-justification. Not only are we special, so are our actions and those of our ancestors. We do not act immorally, so we owe nothing to anyone. False historical narratives are like self-deceptions at the group level, insofar as many people believe the same falsehood. If a great majority of the population can

be raised on the same false narrative, you have a powerful force available to achieve group unity.[33]

Our perceptions of our collective identities are often highly inflated as well. For example, many Americans hold the belief that the United States is the greatest nation on Earth; it's the land of the free, the home of the brave; America is the most prosperous country; Americans are the most generous people in the world; all Americans are created equal; and the American dream is alive and well.[34] Try to put the word "not" in front of any of these qualifications and see how people react. (Warning: never try this if you're a politician.[35])

Such communal aggrandizement or group narcissism is common in other nations as well—such as Russia, China, Germany, and Japan. To borrow from Swiss psychoanalyst Carl Jung, group narcissism is underpinned by the collective unconscious that is endemic to all cultures, whether ancient or modern, tribal or industrial. This is a reason that, for any culture, the view of an insider can be radically different from that of an outsider. That's why anthropologists tend to embrace both views— external (etic) and internal (emic)—to avoid biases and distortions.

ᘐ

Self-deception can reinforce delusional effects, which may in turn lead to superstition. This results because our cognitive responses are biased from having evolved to adapt to our environment. As discussed in chapter 3, when we hear a rustle in grass, our first reaction is a snake—not only a snake, but a venomous snake. Although the reality of a deadly snake in the grass is highly unlikely[36]—unless you're in Australia—our fearful reaction stems from the fact that the cost of a false positive (death) far exceeds the cost of a false negative (a laughable overreaction).[37]

For the same reason, our brains evolved to often trick ourselves into seeing patterns that do not really exist. That's why we sometimes see reports that someone has spotted the image of Jesus Christ in a foamy cup of coffee or on a piece of toast. These are delusional artifacts that result from our brains working too hard to find a pattern when none exists, or to make a causal connection between random events.[38]

Knowing how easily self-deception can lead to delusion and superstition, we shouldn't be surprised when people yell, "Come on! Come on!" to slot machines in casinos, hoping their exhortations to machines might result in a jackpot. Even pigeons are known to engage in ritualistic dances when pecking at buttons on a Skinner box to receive an uncertain reward. Chimps have been reported to perform superstitious rituals such as a rain-dance, indicating a potential evolutionary link between humans and other animals in regard to superstitious beliefs.

But we shouldn't brush off superstition too quickly. It evolved for a major benefit: self-healing. Praying to gods for a cure to a disease is practiced in all cultures. Sometimes, "miracles" do happen through the intervention of a mythic "divine power." Perhaps more familiar to those of us who live in industrialized societies are placebos, widely used to reduce physical or mental symptoms.

One of my memorable encounters with the placebo effect was when I was a high school freshman in 1981 in my hometown village in China. My aunt showed me a bottle of liquid given to her by a sailor from Europe with the word "Derby" in its brand. She told me that the substance was quite potent. Drinking a spoonful of it could cure a range of conditions including cough, stomachache, diarrhea, headache, and heatstroke, among many others—or so the village folk maintained. Since nobody around knew English, she hoped I could tell her what the bottle contained. Unfortunately, my English was too limited to read the label. The magic effects of this "Derby" liquid continued among locals until it ran out.[39]

More than an inert substance like my aunt's mysterious liquid, "a placebo is the whole ritual of the therapeutic act" that convinces the patient a helpful treatment is on the way.[40] Although placebos can generate beneficial effects in a variety of medical conditions and body systems, they are all rooted in psychological effects in general, and self-deception in particular. Just as witnesses' memories are vulnerable to lawyers' suggestive maneuvering in court, our mental state can be affected by placebos through psychological processes such as Pavlovian conditioning, social learning, memory, and motivation. This may in turn activate the genetic, immunological, and neural responses that are unleashed by real drugs with actual biological effects.[41] That's why inactive substances, words, rituals, signs, symbols, or treatments, if perceived as beneficial to our condition, can trigger a placebo effect.

Placebos are known for improving a range of conditions, including sleep, mood, a variety of diseases, and sex lives. Pain researchers have found that one of the primary pathways through which placebos work is kindling hope, which in turn can reduce anxiety. This activates the dopamine-mediated reward center in the brain, reducing a patient's pain.[42] When used with an effective drug (such as the painkiller remifentanil), placebos may further enhance the drug's potency.

The placebo effect can be so strong that at times it may be stronger than the effect of the active ingredients in medicines. For example, antidepressants account for about 25% of improvement in patients, whereas the placebo effect (including spontaneous remission) accounts for 75%.[43] Intriguingly, placebos may still work even when patients are informed that they are being treated with a placebo or a sham procedure.[44]

Not all people respond to placebos, however; nor is it easy to predict who will. Nevertheless, a general pattern is still noticeable for those who do respond: the more expensive the placebo is in terms of the pill or the treatment, or the more invasive the treatment is, the more effective it becomes. The same can be said for the perceived sense of medical authority in therapy.[45] Even the look of a placebo capsule plays a significant role. Dark pills are comparatively more effective for treating pain; warm and bright color pills for stimulants; blue pills for sleep, and green pills for calm.[46] Why?

Associative learning plays a key role in the placebo effect. Besides the color, shape, and taste of pills, other factors such as the appearance of clinics, medical instruments, and health care paraphernalia, and the presence of, or interaction with, physicians and nurses, all have the potential to produce or enhance a placebo response. These settings can mentally suggest to patients that a cure is coming.[47] And the more patients have been exposed to these reminders of the medical profession, the stronger the placebo response will be. Patients, in concert with their immune systems, have been conditioned to respond in this way.[48] This may explain why patients, even after being told that they are being treated with a placebo, may still respond positively.

How can our brains transform otherwise useless placebos into substances that can heal the body? A group of researchers tried to answer this question by inflicting pain on volunteers with either shocks or heat on the forearm and then treating them with an inert cream. As expected, the participants felt markedly better. Brain scans showed that the painkilling effect was real. The brain areas related to pain reduction, known as the "pain matrix" (a complex network including insula, thalamus, anterior cingulate, and other parts), were active after treatment with the placebo.[49]

Furthermore, placebos can increase the activity of neurotransmitters such as dopamine and endorphin in the nucleus accumbens, a specific small structure in the brain. Interestingly, if you use money as a reward to activate this system, the placebo painkilling effect will also increase. And, as you can expect, the more money, the greater the placebo effect.[50] Money indeed can heal! Since learning involves the prefrontal cortex, when people lose control of that part of the brain, they may also cease to respond to placebos.[51]

The entire industry of alternative medicine and healing practices— such as herbal medicine, acupuncture, meditation, chiropractic, and aromatherapy—are, by and large, based on the placebo effect.[52] Acupuncture, for example, is among the best known and most widely practiced alternative healing methods. Chinese practitioners of acupuncture have been plying their trade for three millennia. It's difficult to dispute that it works. But it has remained a mystery as to *why* it works.

After more than 3,500 clinical trials, acupuncture's actual medical effect remains unclear. As such, it is best used for its placebo effect.[53] Indeed, in rigorous trials with randomized control between patients treated with real acupuncture (when the needle is used on the "correct" meridian points) or with fake (that is, putting the needles intentionally on "incorrect" spots) for migraine and chronic pain, researchers found no statistical difference between the real treatment and the fake control.

Moreover, patients' response to acupuncture treatment is typical of a placebo: the higher their expectations, the stronger the painkilling effect.[54] The same results hold for a range of conditions including migraine, tension headaches, chronic low back pain, and knee osteoarthritis.[55] Clearly, the effect is from insertion of the needle itself, not through the meridian (or *jing-luo*) system, as assumed in traditional Chinese medicine. Looking back from our modern perspective, acupuncture may be the best placebo ever invented to trick human psychology: it is invasive enough for people to put their faith in its magic without risking any unwanted side effect.

Some scientists adamantly discredit alternative medical treatments, equating them with snake oil. The hypothetical meridian system that underlies acupuncture has been labeled "prescientific gobbledygook."[56] Such views are biased. After all, medicine is both a scientific endeavor and a healing practice. As a science, medical research must find out whether and why a drug or a treatment works by rigorously following scientific procedures in clinical trials. But as a practice to cure disease, heal wounds, and alleviate symptoms, the primary concern of medicine is *whether* a certain medicine or treatment works, not *why* it works.

It's in this practical sense that alternative medicines should be given their rightful place. Even though they may primarily be placebos and may not cure disease in most cases,[57] they can be useful in accelerating healing or providing relief for those who respond. Thus, they can play an essential role when there is not an effective drug or treatment available. Placebos, therefore, should not be viewed as bogus.[58] Wouldn't it be foolish to leave an evolved mechanism untapped if it can benefit healing, simply because we don't fully understand how it works just yet?

Whether an alternative medicine or healing practice, such as a certain herbal remedy in traditional Chinese medicine, is *useful* can often elicit a spirited debate. The word "useful" can generate a great deal of confusion. In popular usage, it means something that works, as in improve conditions, reduce symptoms, or speed up the healing process. By that definition, placebos are useful, without any doubt. In science, however, "useful" means a positive effect *in addition to* that shown by the placebo effect. By that definition, a drug is considered useless unless it shows an

familiar words such as Napoleon or the Divine Comedy.[63] Similarly, people who feel they have a good grasp of finance are more likely to claim they know about nonexistent financial concepts, and people who feel they are good at geography are more likely to claim they know about nonexistent geographic locations.[64]

As Darwin noted in 1871, "Ignorance more frequently begets confidence than does knowledge."[65] Overconfidence can make people fail to recognize their own weaknesses, leading to narcissistic claims of knowledge they don't really possess.[66] For instance, most people think they know how locks, helicopters, or flush toilets work. But if you ask them to explain, not many can.[67] Likewise, people often think they understand basic financial concepts such as inflation and interest rates. But if you test their knowledge with simple questions, only 56% of people can answer correctly.[68] For the same reason, fans of many sports teams—such as Chelsea FC, the New York Yankees, or the Houston Rockets—claim they can do a better job than the coaches.

Lack of knowledge about how toilets flush poses no real problem in most cases. However, when knowledge is necessary to perform one's job, it's a more serious thing. For instance, professionals such as medical personnel, financial advisors, and legal consultants may not know enough to handle the issues that come up in their jobs, causing physical injury or financial losses to their clients and customers. For example, nearly 40% of first-year medical students fail to perform a CPR exercise adequately, but less than 3% believe they fall short.[69] Would that concern you if your life depends on their knowledge and skills?

Overconfidence, writes Trivers, is "one of the oldest and most dangerous forms of self-deception—both in our personal lives and in global decisions, such as going to war."[70] While it may have been a problem for even our Stone-Age ancestors when decisions affected only individuals or tribes, it can lead to major disasters in modern societies. The *Titanic*, for example, was considered unsinkable by the captain and many others—until the unthinkable happened. The same can be said for many airplane crashes. Worst of all are epic defeats and colossal losses in wars—Napoleon's campaign in Russia, Germany in two world wars, Japan in World War II, the US in Vietnam and in the second Gulf War, and Putin's Russian invasion of Ukraine, to name just a few of the most well-known. All of them stemmed from overconfidence in their own country's military might and downplaying the enemy's strengths and tenacity.

In modern societies, large-scale disasters typically originate from errors and poor judgments in leadership. Studies show that rank, age, and experience are no guarantee of superior performance. (Remember the Peter Principle mentioned in the previous chapter?) Yet those qualities can inspire confidence nonetheless.[71] That's why leaders are prone to be overconfident. Narcissistic CEOs, for example, are particularly vulnerable to self-aggrandizement. When praised by the media or showered with

FIGURE 7.3. Apple CEO Steve Jobs. In adult men, the ratio of facial width to height is positively related to testosterone level (see Lefevre, Lewis, Perrett, and Penke 2013; photo credit: segagman with a CC BY 2.0 license with modifications according to Kamiya et al. 2016).

awards, they can be emboldened to take higher levels of risk while ignoring objective measures of performance.[72]

What leads to overconfidence? Hormones play a part, to nobody's surprise. Testosterone is known to boost confidence and risk-taking behavior.[73] That's why men are more likely to be brash and cocky in comparison with women. Men are also far more likely to engage in thrill-seeking activities such as speeding, gambling, using recreational drugs, and participating in dangerous sports such as skydiving and bungee-jumping. As a result, more men die in accidents (nearly true for me, to add a statistical point) or end up in jail. Likewise, CEOs with higher levels of testosterone are more inclined to take risks in operating their businesses (fig. 7.3). As a consequence, their company's stock prices tend to be more volatile.[74] And just for the record, men trade more frequently in the stock market and have poorer performance, statistically speaking, than women.

Besides business and organization leaders, overconfidence and self-aggrandizement can be detrimental to almost anyone. These derivatives of self-deception can make us close-minded and unable to spot our own mistakes and weaknesses, while resisting good advice and better-informed opinions. For instance, smokers are less willing to accept information about the danger of smoking than nonsmokers. Some people avoid taking HIV tests with the attitude: "What I don't know can't hurt me."[75]

𝕏

Self-deception can be reinforced by a cognitive loophole: confirmation bias, a psychological term for our preference for ideas and facts that agree with our worldviews, while avoiding or filtering out information that contradicts our beliefs. This bias can feed our self-deception by protecting our self-esteem, pride, and ego. Even when the facts show otherwise, some still hold onto the hope that the evidence may be false.

Confirmation bias is a cognitive weak spot that can be exploited on a grand scale. There were few people who rivaled Frank Abagnale's exploits until the "Cryptoqueen," a 36-year-old woman by the name of Ruja Ignatova (fig. 7.4), popped into view in June 2016 in London, as reported in a stunning story by the BBC.[76] Known in public as Dr. Ruja, she took the stage at the most prestigious venue for tennis, Wembley Arena, and announced to the world that a new cryptocurrency, One-Coin, would be "the Bitcoin Killer." "In two years," she told a frenzied crowd, "nobody will speak about Bitcoin anymore."

Like all money-crazed cults, people believed in her and followed her as she spoke to excited crowds all over the world. In less than three years, from 2014 to 2017, people bought over $4.5 billion worth of One-Coin. They poured their hard-earned money into packages sold by her company and happily calculated their wealth as they saw the "value" of OneCoin creeping up. They not only bought the cryptocurrency themselves but also dragged their friends and relatives into this "once-in-a-lifetime" opportunity. Dr. Ruja's story was so compelling that even a pyramid scammer, himself rebranded as a multilevel marketing guru,[77] was convinced to put tens of millions of dollars in OneCoin, thinking his wealth could very soon top Bill Gates's.

FIGURE 7.4. "Dr. Ruja"; Ruja Ignatova (photo credit: Onecoincorporation with a CC BY-SA 2.0 license, no modification made).

But none of those things would ever come to pass. The real Dr. Ruja was a stay-at-home mom from Germany. Her elaborate operations were run by a shadowy company headquartered in an apartment complex in Sofia, the capital of Bulgaria. Her credentials as a leading businesswoman came from a paid advertisement on the back cover of the Bulgarian edition of the magazine *The Economist*. And most bizarre of all, the highly touted cryptocurrency OneCoin never even existed.

Bjorn Bjercke, a Norwegian blockchain expert, smelled something fishy in the quick-money frenzy. He reached out to some of the victims, warning them about the scam. To his great surprise, his effort to expose the fraud often got him in shouting contests with the victims, who refused to believe what he said. He even received death threats, apparently not only from those who profited from the scheme but also from some of the victims. Indeed, how could people concede their loss and move on after putting in so much money, along with their belief, passion, reputation, and pride? The company behind OneCoin knew people's weaknesses well and urged stakeholders-cum-victims to shout down skeptics and troll critics of OneCoin. This was particularly dishearten-ing for Bjercke, who thought he was doing society a favor. "If I knew what I would have to go through," he said in an interview, "I would have never blown the whistle."

Not until October 2017, when Dr. Ruja failed to show up in Lisbon for a hyped public speech, did people realize it was all a big lie. She has since become a fugitive. On November 5, 2019, her brother, Konstantin Ignatov, confirmed the fraud while testifying in court in New York, claim-ing that even he had been conned by his sister. The BBC report ends:

> Dr Ruja identified several of society's weak spots and exploited them. She knew there would be enough people either desperate enough, or greedy enough, or confused enough to take a bet on OneCoin. She understood that truth and lies are getting harder to tell apart when there is so much contradictory information online. She spotted that society's defence against OneCoin—the law-makers, the police, and also us in the media would struggle to understand what was happening.

If the OneCoin story tells us anything, it's that in the Information Age, information can hurt us—unless you know its accuracy and reliability. And confirmation bias—an otherwise subtle, innocent preference for what we like to hear—has emerged as a major liability. Indeed, the One-Coin scam was so spectacularly successful because of its methodical exploitation of people's confirmation bias. It encouraged them to cherry-pick false information that appealed to their preferences while rejecting truthful information that would otherwise protect them from deceit.

Confirmation bias can convince us to become willingly trapped in a cozy, self-made cocoon, isolated from objective reality. It can drive us to the most extreme manifestations of the Dunning-Kruger effect—to the point where we may no longer recognize ourselves. Today, the Internet, powered by AI and Big Data, knows you perhaps even better than you know yourself: your age, gender, education, hobbies, marital status, political leanings, artistic taste, and far more—including private information such as financial status, personal obsessions, and sexual fantasies—that you wouldn't share even with your closest confidant.

Amazon.com knows what you want to buy; Netflix, what movies and shows you like; Google, what you like to read and watch. That's why you get recommendations for news articles, videos, movies, things to buy, and people to date. When you spend money, they profit. These individual-targeted marketing strategies are, in principle, similar to the OneCoin scam in the sense that both are designed to exploit our cognitive biases—that is, the Second Law. They are legal because they are not lying. The major difference between them is that in addition to the Second Law, OneCoin also exploited the First Law by lying, which was illegal.

Joseph Goebbels, the Nazi Reich Minister of Propaganda, is often credited with the claim, "Repeat a lie often enough and it becomes the truth." As we know, simple repetition of the same lie (that is, using the First Law) may not be such an effective technique. To brainwash the masses, you need to resort to the Second Law, too, by exploiting cognitive biases to the maximum degree possible to make people follow you willingly, blindly, and fervently. Profit motive aside, many feel-good ideologies—such as prosperity, patriotism, freedom, liberation, and national strength—can spark a nationalist passion by feeding and pumping up self-deception among the populace of a nation. When a critical mass of people believes something is true, it becomes true in a very real sense. (Think about QAnon as a case in point.) But it all begins with the echo-chamber effect built on the cognitive loophole of the confirmation bias.

One of the most stubborn echo-chamber effects in our time belongs to the false belief that vaccines can cause autism. It all started with a single fraudulent paper by Andrew Wakefield in the British medical journal *The Lancet* in 1988. There is still a small but vocal group of people who continue to cling to the false belief today, even years after Wakefield was barred from medical practice for fudging results in the paper for financial gains.

Echo-chambers close our minds, encouraging us to associate only with those who cling to the same ideas, even when they are false and harmful. As such, they are fortresses walling out information that disagrees with our preconceived opinions. You know how hard it is to sway the minds of friends or relatives when their views are different from yours on such hot-button issues as abortion, immigration, climate change, gun control, or the death penalty.

Unlike its physical auditory analog, the informational echo-chamber unfortunately does not keep "echoes" within its confined space. People spread words and ideas, which may allow a relatively small sound in one echo-chamber to be amplified to a loud and pervasive noise in society at large. Indeed, a recent study shows that a small number of people have an outsized impact on the proliferation of fake news. Tweets containing false information, when designed to be novel, catchy, and surprising, can travel six times faster than truthful tweets. Furthermore, they are more likely to go viral, spreading to far more than 10,000 Twitter users. Truthful tweets, in comparison, rarely reach more than 1,000 people.[78] This was exactly what happened during the 2016 election. Nearly 80% of the total shares of fake news originated from a mere 0.1% of people—largely old, conservative men who were interested in politics.[79] More amazingly, a 2021 analysis of Twitter and Facebook by the Center for Countering Digital Hate shows that 65% of disinformation about COVID-19 vaccines initially came from a mere 12 anti-vaxxers. And on top of the "Disinformation Dozen" is Joseph Mercola, a Florida osteopathic physician who reportedly made millions of dollars from selling natural health products (such as vitamin supplements) claimed to be alternatives to vaccines.[80]

Confirmation bias can become a major obstacle for constructive discussion among people holding different opinions. A recent study shows that even when a news outlet issues a correction for a previously published falsehood, those who believe it still refuse to change their views.[81] Worse, showing either liberals or conservatives a view that differs from their own will only cause them to dig in their heels on their own position.[82] Even during times of crisis when citizens are more motivated to seek information that will lead to real solutions, they may still choose information that supports their own preconceived ideas and ignore information that doesn't.[83] Apparently, denying reality is a coping mechanism to protect people's psychological well-being during times of stress, terror, and tragedy.[84]

Such a mass mentality is often exploited by partisan ploys. Many political strategists are professionals who design campaign strategies that make maximal use of voters' cognitive loopholes—especially confirmation biases—on social, economic, cultural, and military issues. That's why every issue that can be politicized—including GMOs, climate change, immigration law, gun control, and even whether to use masks to fight COVID-19—has been politicized to the extreme. It's no wonder that political debates are so often little more than shouting matches devoid of substance and productivity.

Partisanship is further promoted by media companies that take advantage of our confirmation biases for profit. One news outlet explicitly advertises itself by promising to provide its audience with "news we can agree on." Some cable news channels fill their airtime with pundits espousing provocative views, extreme opinions, and groundless conspiracy theories to pander to and influence viewers, instead of providing actual news. Some even spin entirely false narratives that have nothing to do with the truth. Instead of practicing journalism, their only concern is ratings because that is what brings them money from sponsors and advertisers. When a media company openly calls a political opponent "the enemy of the people" or the COVID-19 pandemic "a hoax," how much more will it take to become a full-blown propaganda machine?

𝕬

As Confucius noted 25 centuries ago, quoted in the chapter epigraph, ignorance about oneself is common; this has baffled philosophers and thinkers ever since. "What indeed, does man know of himself!" Nietzsche exclaims in one of his works, *On Truth and Lie*, and then poses a string of questions:

> Can he even once perceive himself completely, laid out as if in an illuminated glass case? Does not nature keep much the most from him, even about his body, to spellbind and confine him in a proud, deceptive consciousness, far from the coils of the intestines, the quick current of the blood stream, and the involved tremors of the fibers?

If no one was able to answer all of these serious questions in Nietzsche's era, we can do a better job now with knowledge of the Dunning-Kruger effect. This is why the worse people perform, the better they feel about their performance; the less they know, the more confident they become; the more insecure they are, the more likely they are to reject information that contradicts their views. These can all lock people into traps of their own making. Thus, self-deception is a major roadblock to self-improvement. That's why Richard Feynman warned college graduates in his commencement speech at Caltech in 1974, "The first principle is that you must not fool yourself—and you are the easiest person to fool." Nicely said! But how to do it?

Fortunately, the answer is also in Dunning and Kruger's original study, shown in figure 7.1: outperformers tend to underrate themselves, a result that has been confirmed in other similar studies as well. For example, people who feel more secure are more willing to absorb information that disagrees with their views.[85] This in turn can help them make progress in many aspects of their lives, leading to a positive feedback loop in the other direction. That is, modesty and humbleness will make us better, whereas pride and vainglorious perceptions do nothing but make us retreat into narcissistic fantasies and self-constructed shells.

So, the wise become wiser *because* they remain humble and self-critical. That's why sages such as Confucius, Socrates, Darwin, Einstein, and many others are famous for their modesty, which motivates them to learn from mistakes and overcome their weaknesses. There is no obvious reason why we can't do as they do. The question is how.

The most obvious approach is to follow the advice of wise and successful people. Here are some famous quotes that can be used as mottos or reminders. Are you familiar with any of them? See the answers in the endnote:

1. There is nothing noble in being superior to your fellow man; true nobility is being superior to your former self.
2. It is unwise to be too sure of one's own wisdom. It is healthy to be reminded that the strongest might weaken and the wisest might err.

3. A great man is always willing to be little.
4. A true genius admits that he/she knows nothing.
5. I have three precious things which I hold fast and prize. The first is gentleness; the second is frugality; the third is humility, which keeps me from putting myself before others.
6. A man should never be ashamed to own that he has been in the wrong, which is but saying in other words that he is wiser today than he was yesterday.
7. A man wrapped up in himself makes a very small bundle.
8. Humility is the solid foundation of all virtues.[86]

Also, as the gap between self-assessment and peer-assessment tells us, outside input can give us a valuable dose of reality. We would all benefit by regularly asking others for their honest feedback about ourselves. Meanwhile, we should also offer help to those who are trapped in the vortex of the Dunning-Kruger effect. However, since underperformers are more likely to ignore criticism, we need to be savvy in our approach. Smokers, for example, tend to be defensive if reminded of the downsides of smoking, such as wasting money or getting cancer. But if they are praised for their kindness, they can become more open to anti-smoking campaigns.[87] The bottom-line message is that we should not harm people's self-esteem while helping them overcome self-deception.[88]

Who can we seek help from besides sages who are mostly deceased? Well, one possibility is to tap into a living, breathing group of people who are statistically better grounded in objective reality: women. Women are more likely to play down their own abilities than men, sometimes to a degree that they may fail to see their own strengths.[89] One study shows that, compared with men, women underestimated themselves by 13% in the number of questions answered correctly and by 17 percentile points lower in performance.[90] Women's apparent self-doubt is often mistaken as lack of confidence. It has been stereotyped in traditional Eastern and Western societies as a gender-specific weakness, molding the popular but false belief that women are unsuited for leadership. (Remember the famous line, "Frailty, thy name is woman!" in Shakespeare's *Hamlet*?)

Such a prejudiced view couldn't be farther from the truth. Admitting a little self-doubt gives women a better grip on reality—they are markedly less likely to fall victim to the Dunning-Kruger effect. The qualities we admire in sages—modesty, humbleness, and skepticism—enable women to better perceive objective reality, essential for making sound decisions. Those qualities should be taken as strengths, rather than weaknesses, in leadership.

This theory has been supported by data. In the early 1990s, California began to encourage recruiting women as board members in public companies. It was originally meant to promote gender equality. Some were skeptical, whereas others even resented the idea as only playing gender politics. But after a while, people began to realize that companies with women on their boards tend to perform better financially. Call it the mystery of women board members. Is this just a fluke?

The answer is no, according to a recent meta-analysis relating types of leadership roles to a company's financial performance in eleven metrics. While the study confirms the positive impact of women's leadership in general, it points to two specific areas where women clearly outperform men. One is sales, and the other is the board of directors.[91] Women CEOs, however, fail to outshine their male counterparts. Why? Apparently, women's leadership strengths stand out most in making collective decisions.[92] CEOs, however, often call the shots all by themselves. Although women are just as competent as men in these leadership positions, their unique advantage in a group scenario disappears. So, the mystery of women board members is solved.

Although women's leadership doesn't underperform men's in any category, don't rush to assume that women leaders could alone make your company better off financially. A statistical pattern does not speak for individual companies. If your company is Hewlett Packard, for example, you would regret having hired Carly Fiorina as CEO, the first woman to head a top tech company in 1999. She made so many wrong calls on key decisions that her leadership resulted in the company losing half its value. The pain didn't end until she was ousted in 2005.

Moreover, there is no evidence demonstrating that women do better on an all-women's team. Maybe, confidence is still an important factor in getting things done with speed and determination in business. It appears that a mixture of men and women in leadership may provide the best outcome statistically. That way, you can have enough confidence but not overconfidence. Unfortunately, today only 4% of CEOs and 16% of board members in Fortune 500 companies are women,[93] indicating that we are still far from making full use of women's psychological advantage.

At this point, we have paraded through vast varieties of cheating in a wide range of organisms and discovered that all forms of cheating are done through the use of the "two laws"—falsifying honest messages in communication and exploiting cognitive loopholes, which are biological foundations for lying and deceiving, respectively. We have also tried to fit human cheating into the big picture of the biological world with emphasis on its diversity, complexity, and uniqueness. Before we wrap up, we have yet to answer one more major question: What can we do about cheating?

CHAPTER 8

Living with Lies and Deceptions

Lying is universal—we all do it; we all must do it.

—MARK TWAIN

Seattle's famed Pike Place Market never fails to attract crowds. On seafood stands, halibut fillets glisten and crabs appear to wiggle their claws. To add some seafaring spirit and fun, vendors mimic fishermen's chants when they make a sale.

On a crisp Saturday morning, you are part of the bustling crowd. A perfect-looking king salmon catches your eye. It's so appetizing that you feel you would miss the culinary experience of a lifetime if you can't have it. You ask the vendor for the price.

"$3.50 a pound. Oh, I haven't put out the price tag yet." He smiles, his eyes inviting.

"What?" You can't believe what you've just heard. "It sells for $12.99 a pound in town, and the quality is not even close."

"You see," the man disagrees, "it has already been four hours since the fish was landed." He points at a flaw on the tail. "And some scales are missing here. If I can make 25 cents a pound, that's more than enough for me to make a basic living."

"He's so honest," you say to yourself. "I could never shortchange such a nice guy." So, you insist on paying $12.99 a pound, but he adamantly rejects your generous offer. Since you can't come to terms on price, the deal is off.

Feel strange? This is a world where people do business with all honesty, all other-concern, and no attempts at deception. (Of course, the real Pike Place Market would never exist in such a business utopia.)

As much as we loathe lies and deceptions, this thought experiment makes a case for the necessity of cheating in business dealings. In this chapter, we will see that cheating, contrary to its pervasive notoriety, is actually essential in our economic activities and social lives, and it is a critical part of our education and cognitive development. So, there is no question as to whether cheating is permissible. The question is what kind of cheating should be allowed and when it is morally justifiable. And more pragmatically, how can we live—thrive, even—in a world full of lies and deceptions. Here, we will search for answers to these hard questions.

ℱ

The market only knows profit, not conscience. "It is not from the benevolence of the butcher, the brewer, or the baker that we expect our dinner," Adam Smith wrote in *The Wealth of Nations*, "but from their regard to their own interest." Without a profit motive, there would be no marketplace, nor any of the economic activity that surrounds it. We would be back to the Stone Age, at least in an economic sense. This is what could happen if Adam Smith's famous "invisible hand"—the incentive for profit—were to be chopped off.

Although ECON 101 tells us that price depends on supply and demand, bargaining allows sellers to seek the highest profit and buyers to pay the lowest price possible. It is, in essence, a strategic game played between the two parties to resolve their perennial conflicts of interest. Unsurprisingly, bargaining and negotiation are rife with such tactical tricks as subtle bluffing and blatant lying to get a better price. In fact, bargaining and negotiation are potential breeding grounds for unethical behavior and are fraught with deceptive maneuvering.[1] Nevertheless, few see the bargaining process as intrinsically unfair, even for buyers who feel they were ripped off. People are free to haggle over prices in any way they want; that's why it's called a "free market."

Bargaining is a vital process in cutting deals. It is a mental game in which one party tries to outwit the other by using whatever means necessary. As such, being shrewd is not only desirable but also essential. A poker face, like in that bluffing-based card game, is required if you want to win. Would you accuse a bluffing poker player of being dishonest? Likewise, when Google moved to acquire YouTube in 2006, both sides kept their price tags under wraps during the negotiation. Had Google officers honestly told YouTube what Google knew and the price they'd offer, they would have faced criminal charges of industrial espionage for theft of trade secrets.[2] In this case, honesty is not just a liability—it's illegal.

If you dislike bargaining and prefer seeing a price tag instead, you lose the opportunity to negotiate for a better price. You would have to accept whatever the posted price is—if, for instance, salmon sold at $3 a pound at a fishing dock or at $100 a pound at an upscale restaurant. Consider the following extreme case. When the pharmaceutical company Novartis brought to market the anti-cancer drug Kymriah in September 2018, the price tag was $371,000 per treatment. It stirred up a flood of public outcry, accusing the drugmaker of being dishonest, unfair, and greedy. But Novartis could easily fend off these charges by claiming that developing the immunotherapy drug cost them billions of dollars and they needed to make a profit to survive. Who deserves the blame—the company or the economic inequality in society?

Even in a supermarket where everything is tagged with a price, deception still takes place because items are often deliberately arranged to exploit your cognitive biases. Stores are designed to appeal to your senses (including ambient lighting, music, and aromas) so that you can "enjoy your shopping experience"—a thinly-veiled synonym for "open your wallet." Shelving, in particular, is a subtle maneuver employed by many stores where the items displayed near the checkout area may be considerably more expensive than the same items on regular shelves. You pay for the "convenience" when waiting to check out. Cereal boxes may be displayed on lower shelves, instead of at eye level, because children can see them better and can twist their moms' arms to get what they want. Knowing the psychological impact of shelving, manufacturers may negotiate with retailers as to where their products get placed.

Marketing, in a nutshell, is about psychology—often taking advantage of consumers' cognitive biases, including framing, anchoring, and loss aversion, just to name a few on a long list. The examples of in-store marketing used here are collectively known as nudges, a word coined by behavioral economist Richard Thaler and law scholar Cass Sunstein.[3] Nudges are cognitive maneuvers aimed at channeling our state of mind into making decisions desired by the manipulators. For this, marketers are keenly interested in cutting-edge neuroimaging research, with the intention of designing and advertising their products by appealing to our patterns of brain response.

Marketing campaigns that are intended to trick our senses or exploit our cognitive biases are clearly deceptions. Even so, we tend to accept them. Do we really have any control over how marketing is practiced? Or, does cheating even matter here?

Human behavior is subject to two control forces: legal and moral. However, these two forces do not necessarily lead to the same result. People can behave in ways that are completely legal yet totally unethical at the same time. A completely unregulated free market, as Adam Smith saw it, is amoral. Profit motivates producers and sellers to bring goods and services to the market. It also gives them the incentive to resort to deceptive tricks. That's why few are surprised by the prevalence of strategic cheating in the marketplace. In fact, the more that profit is at stake, the more likely it is that deceptive ploys will be used in bargaining.[4] The Chinese even have a spiteful saying for those involved in trading—"no merchants are trustworthy." Even though you might not buy into such a cynical perspective, it would be naïve to assume all people in the market behave honestly with benevolence and a strong moral sense.

Indeed, the market only knows what is legal or illegal. Morality plays no role. Economist Milton Friedman embraced laissez-faire capitalism, claiming that a firm's responsibility is "to make as much money as possible while conforming to [the] basic rules of the society, both those embodied in the law and those embodied in ethical custom."[5] "Ethical custom" is used here in the normative sense, referring to conventions practiced in an industry, often without binding rules. This justifies deceptive tricks as part of the game of doing business, as long as they stay within the bounds of the law.

Although ethics in business and marketing has become more of an issue in recent decades, it hasn't extended to practices in pricing or bargaining. Currently, marketing ethics deals with truthfulness in listing product ingredients and transparency about their potential risks concerning health, environmental impacts, financial concerns, privacy, and security issues for consumers.[6] Advertisers, however, are only required to provide truthful information; that is, they can't lie (cheat by using the First Law). However, that still leaves the door open for advertisers to use deceptive maneuvers (cheat by using the Second Law). Most controversial marketing strategies (some even illegal in some nations)— shilling, bait and switch, pyramid schemes, viral marketing—fall into the second category. They exploit consumers' cognitive loopholes by using inflated language and alluring designs, or coercing people to consume by cultivating and fueling a cultural fad.[7]

The bottom line is that sellers never use (honest) negative language to describe their products or services. They would never say, for instance, "this product contains 5% bad fat," or "this investment opportunity may bankrupt you." Only when forced to do so by law do manufacturers add warning labels (such as those on all tobacco products) to alert consumers about potential risks. Even so, they try to use the smallest font possible and bury it in an obscure place on the packaging to ensure that only the most careful consumers will ever read it. Therefore, honesty and integrity are relative in the business world—relative, that is, in terms of existing norms and standard practices.

The above scenarios lead us to a moral terra incognita. They force us to confront several pragmatic dilemmas: Can we live in a society without cheating? Are certain lies and deceptions allowed? If so, what kinds? And finally, how can we live with and make the best out of cheating?

訛

Wouldn't it be nice, you may be wondering, if we could all be honest all the time? Not necessarily. The Chinese social critic Lu Xun, in his short essay, *Proposition*, written in 1925, tells a story about a family showing their newborn boy around the neighborhood, celebrating the 100th day of his existence. During the happy event, one guy in the gathering declares, "The boy will strike a fortune in the future." He is thanked profusely. A second lad chimes in, "The boy will become a high official in the future." He too receives a great deal of gratitude. A third fellow mutters matter-of-factly, "The boy will die in the future." He is instantly disparaged, cursed, and battered.

Of course, either striking it rich or ascending to the heights of government hierarchy is a low-probability event in the boy's future. (This was especially so in China at the time of Lu's writing.) Both are compliments bordering on lies. But death in the future is inevitable, a certainty for everybody. Yet the truth teller is seen as socially awkward and malicious, whereas the two flattering liars are socially rewarded for being well-meaning and kind.

If you find yourself in this situation, what can you do? If you want to be honest without committing social faux pas, the best you can say, according to Lu, is, "Hey, look! The boy is so. . . . Wow! Ha-ha!" But even beating around the bush with an equivocal "wow" or "ha-ha," you are still being dishonest because you are withholding the truth for fear of giving offense. In other words, you are being a hypocrite. The case illustrates that there is no honest and graceful way out in some situations.

This is a common scenario where honesty is punished and lies—well-intentioned white lies though they may be—are rewarded and encouraged, and are deemed socially appropriate. Such is our world today as it was for Lu Xun a century ago. There is no reason to believe it was different at the time of Confucius in the East or of Socrates in the West. Can you imagine a society without this kind of cheating? We tell white lies all the time, saying nice things to make others feel good, even though they are mostly untrue. On the contrary, a person who lacks such social skills—that is, lying—will be seen as rude or awkward.

Lu Xun's story also illustrates a typical example of what I call *the peril of honesty*. Honesty, besides its clear definition in communication, is also a social norm that varies among communities and cultures. In other words, *honesty in society is more about local social customs, conventions, and standards* rather than a fixed system of personal ethical conduct or moral beliefs. In American society, for example, neither Washington (who famously said, "I can't tell a lie.") nor Trump (who claimed, "I will tell the truth if I can.") seems to be quite in sync with the norms practiced by the vast majority of citizens. And neither was Socrates, mentioned in chapter 1. Socrates might have intended to use his life to set an example for promoting trust and honesty in society. In reality, he laid down his life for holding to a *personal* moral standard that was neither accepted nor appreciated by Athenians.[8] He lost his life, and the world lost a great man too soon. A double tragedy, indeed!

The same can be said if you apply for a job by sending out a resume that lists all your weaknesses. Although honest, your resume guarantees your rejection everywhere you apply because you're confusing your personal moral standard with the societal norm. You are selling yourself in a wrongheaded way, not just selling yourself short. You will suffer from *the peril of honesty*. In other words, society has its own mind in judging which standards of honesty we should adhere to and which we should ditch.[9]

As the word "peril" indicates, honesty can be disastrous. Imagine you are a middle-class wife who married a handsome husband 15 years ago. One evening, your husband is about to attend a formal public event. After 20 minutes of careful grooming and trying out suits, he comes to you and asks, "Honey, how do I look?" He is expecting you to say, "Great!" as you would normally do. But that moment, you decide you've had enough with white lies. You summon up your courage and answer him in a burst of honesty, "Oh, just so-so," and add, "You're bald and fat. Our neighbor Jack is fit and athletic. And my boss has just bought a new Corvette." Even though both you and your hubby know full well you are telling the truth, this is not helpful for both him and you. If you are equally "honest" with friends and relatives, you will guarantee your status as a social pariah. You would be lucky if you were simply labeled a misfit or weirdo.

The above scenario would have been hypothetical, but the BBC decided to give it a try with real people in 2018. In the documentary, *A Week without Lying—The Honesty Experiment*, three volunteers—a priest, a YouTuber, and an advertising executive—were recruited to take part in an experiment where they were not allowed to tell any lies, even white ones. Although they seemed happy to sign up for the test, the participants soon discovered how brutal it is to live an entirely honest life. One of them even took sick leave to avoid interacting with others at work.

The BBC experiment illustrates how perilous honesty can be in our social lives. Clearly, without white lies to signal that we are empathetic and supportive and harbor no ill intent, we would be seen as sociopaths who care little about the feelings and well-being of others. This again illustrates that honesty is what society sees in us, rather than ways of telling the whole truth.

The peril of honesty speaks to the necessity of cheating in our social lives. For example, if you go to a law office to seek legal representation and are greeted by a lawyer in a T-shirt and a pair of slippers, are you likely to hire him? What would the judge and jury think if your lawyer wore pajamas in court? The same logic works for business professionals—bankers, financial consultants, or company executives. If they don't somehow draw a distinction between themselves and people in the street, will they be seen as successful? What if an investment guru has long, unkempt hair, wears a pair of ragged jeans, and drives an old clunker of a car? Would these inspire confidence? Would you trust him to manage your life savings?

We often complain that people are shallow and tend to judge a book by its cover. The reality is that we are no longer living in close-knit communities where people know each other by direct reputation. In today's open societies, large cities in particular, people come and go. Dunbar's number demonstrates that we cannot possibly keep track of more than a few hundred people. We must deal with strangers all the time, for a wide variety of reasons. This forces us to make snap judgments because people don't have the time or opportunity to get to know each other. Even so, we often have to make decisions in the here and now. Judging a book by its cover may not ensure accuracy, but when flash decisions are necessary, would you pick a book with a poor cover design if you had another choice? The takeaway message: when nothing else can be used to judge the quality of a book, the cover matters. That's why packaging design is so important in marketing commercial products today, be they books, wines, or shampoos.

First impressions, as a result, become supremely important in our fast-moving society. And whether we are aware or not, much of what we wear and how we appear in public is intended to appeal to others, often at the expense of our time, money, and comfort. This includes our hairstyle, dress, shoes, perfume or cologne, and car. For women, walking around in even moderately high heels can be painful. Using various forms of social camouflage—pretending to be different people than we really are—becomes a necessity for social survival.[10] That's why we have stereotypical images of how professionals should look and behave; that's why we groom ourselves well for dates or public events; that's why we resort to conspicuous consumption to reassure others of the authenticity of our public persona. In most of these cases, we are handicapping ourselves to communicate honest information, an evolutionary principle that we explored in chapter 4.

Even the most famous and wealthy aren't free from pretension. For people like Jeff Bezos, Elon Musk, and the biggest sports or movie stars, it may make no difference whether they wear Armani or Levi's or whether they drive a Mercedes or a Chevy. But they often have something else to hide from public scrutiny: secret affairs, personality weaknesses, unhappy family lives, or questionable business dealings.

As we enhance our public image, we cover up our weaknesses in body, skills, intellect, and socioeconomic status. In both Eastern and Western societies, many women regularly use makeup to improve their natural looks—smoother, more colorful skin, larger eyes, and redder lips. Clothes are designed to enhance a woman's body curvature; tights and jeans to emphasize the lower part of the body, especially legs; belts to accentuate the waist-to-hip ratio. In some highly competitive Asian cities (such as Seoul and Guangzhou), where people are socially compelled to show off their best faces in public, even men are known to wear makeup.[11]

What would our world be like if people didn't care about their public appearance? The logical answer: everyone would dress purely for warmth and comfort. There would be no makeup, no business attire, no wedding dresses, no tuxedos, no manicures, no hairdressing. The fashion and personal-care industries, especially at the high end, would be dead. With style, taste, and vogue all things of the past, our haute couture would lose its raison d'être. Some sorts of lies and deceptions are not only inevitable but essential for our economy and culture.

Cheating is also the backbone of literature. Novels are, by and large, make-believe stories, despite a solid basis in reality for most. So too are memoirs, autobiographies, and narrative nonfictions, most of which are rife with unreal but delicious details to drive the story. Even fact-based nonfiction can contain spurious information. Malcolm Gladwell is among the most popular writers of our time with *The Tipping Point*, *Blink*, and *Outliers* among his best-selling books. Psychologist Christopher Chabris, after finding that Gladwell was cherry-picking the research results he presents, posed a rhetorical question: "Is there no sense of ethics that requires more fidelity to truth?"[12] "Chabris should calm down," Gladwell retorted. "I was simply saying that all writing about social science need not be presented with the formality and precision of the academic world. There is a place for storytelling, in all of its messiness."[13]

Stories and plots—no matter if they are real, semi-real, or totally fabricated—provoke our emotions and subjective experience, without which our intellectual life would be greatly impoverished. That's why we value a wide variety of literary forms—poems, novels, plays—for their creativity and imagination. The same can be said for fairy tales and religious stories. If all writing were restricted to dry facts, we'd be left with nothing but almanacs, chronicles, and academic research articles, none of which qualifies as literature.

Illusion and deception—even intentional misdirection—are key ingredients in artistic works such as paintings, music, movies, TV shows, and video games. These all give us new perspectives on life and exotic experiences through visual images, sound effects, and virtual realities. For example, motion pictures, TV screens, and computer monitors all trick our eyes by connecting a series of still images that recreate continuous movements as if they were real. They have become such an essential part of our lives that without them we might feel uninformed, uncivilized, or at least unsophisticated.

Cheating not only underpins many aspects of our economic, intellectual, artistic, and social lives, it also helps define our desired moral values. Without lies and deceptions, who would care about honesty? Without betrayal, who would feel inspired by an enduring friendship? Without infidelity, who would be moved by a lifelong devotion to love? Without cheating, what would be the point of promoting fidelity, loyalty, decency, respect, reputation, and other virtues that underlie truth and trust?

ƛ

As much as we may hate cheating, we ourselves often resort to it, whether we're aware of it or not. Even Socrates, who sacrificed his life for truth and trust, wasn't disillusioned by reality. "The greatest way to live with honor in this world," he once confessed, "is to be what we pretend to be."

Honor, dignity, and virtue aside, cheating is inescapable for more pragmatic reasons: cheating is a strategic option for both taking advantage of and defending against it.[14] It is in this sense that philosopher and writer David Livingstone Smith calls cheating "the Cinderella of human nature[,] essential to our humanity but disowned by its perpetrators at every turn," in his book *Why We Lie*. He continues:

> It is normal, natural, and pervasive. It is not, as popular opinion would have it, reducible to mental illness or moral failure. Human society is a "network of lies and deceptions" that would collapse under the weight of too much honesty. From the fairy tales our parents told us to the propaganda our governments feed us, human beings spend their lives surrounded by pretense.[15]

Since cheating is such an essential part of our social intelligence, we may not be able to succeed without it in arenas such as government, corporations, schools, and military organizations. In social intercourse, being honest is often far from sufficient to form and sustain human relationships. We need to communicate good intentions quickly and effectively, which often requires positive attitudes, euphemisms, and white lies. It's truly ironic that we have to resort to dishonesty to convey information that convinces others of our sincerity.

Equally important, cheating can hone our cognitive acuity to shield us from falling prey to scams. As we know, a mentally sharp person is more likely to spot a liar by reading facial expressions, conversational styles, and body languages. By contrast, people who have lost some of their cognitive capacity are more vulnerable to swindles and scams. But in general, as research shows, we are not particularly good at detecting cheats—only a tad better than a coin toss. But with training and experience, this ability can be improved. That's why professionals such as law enforcement officers, clinical psychologists, and secret service agents are markedly better at detecting lies and deceptions.[16]

These are some of the reasons why cheating is essential for our mental and social development. Apparently, cheating is an instinct that can show up even without learning. I was shocked to notice that one of my sons fake-cried when he was a mere six months old. Such an anecdote surprises few psychologists, though. They know that two-thirds of children aged two-and-a-half can use at least one deception per hour. At this age, they tell lies mainly to serve themselves, particularly to escape punishment. As they grow up, the nature of cheating begins to shift, from self-serving to other-concerning. By age five, children can already tell white lies to protect others' feelings.[17] A study in China shows that as many as 40% of seven-year-olds, 50% of nine-year-olds, and 60% of eleven-year-olds tell such prosocial lies.[18]

According to psychologist Victoria Talwar and colleagues, children's lying skills come in three stages. At around age two or three, they can make false statements, or primary lies. These lies are mainly to evade punishment for violating rules. At the age of three or four, most children can lie to cover up their wrongdoings. As these lies are used to conceal violations, they are called secondary lies. Finally, at around the age of seven or eight, children become so skilled at lying that they can tell "tertiary lies." That is, they can lie about lies. Thus, they can build a chain of lies that are logically coherent. Now, it's not easy to tell whether they're lying or not.[19] At this point, they're well under way toward learning the ropes and integrating themselves into the adult world, picking up on others' feelings and acquiring skills to deal with complex interpersonal issues.

Cognitive skills such as perspective-taking and mind-reading are not only necessary for successful lying but also for lie-detection. It's unsurprising that children's physical and mental health are positively related to their ability to lie and deceive.[20] As we learned in chapter 5, smarter kids are not only more likely to lie, but they are also more capable of lying and detecting lies, for they have better executive function in their brains—better working memory, cognitive flexibility, and inhibitory control.[21] Autistic children, however, often have difficulties in lying and detecting lies.[22]

As essential as they are for social survival, skills in cheating and cheat detection take prolonged and extensive learning in children. Apparently, many childhood games are designed for this purpose, such as peekaboo, hide-and-seek, card tricks, magic, and liar's dice. Some are intentionally used for children to learn and hone their deception-spotting skills, such as teasing, pretending, concealing, and distracting.[23] Lying also helps children develop linguistic skills in expressing and discerning meaning, tone, and social context of words and sentences. For this reason, some researchers believe that lying has a significant impact on the increasing sophistication of language over time.[24]

In modern societies, children are helped by parents, caregivers, and teachers to develop skills in cheating and cheat detection, necessary for a successful transition to living in the adult world. As early as they can reason, most children begin to be taught when and how to lie. Mostly, they're encouraged to tell prosocial lies and are disciplined for telling antisocial lies. As children learn what kind of cheating is permissible and what is not, they may gradually establish a moral sense of right and wrong. Conversely, children who fail to master these social and cognitive skills and learn these ethical norms may exhibit maladjusted behaviors unsuited for the adult world.[25]

⚶

Cheating is as inevitable as it is essential for living in human societies. But as the most intelligent primate, we humans think—and think big—about what we *ought to* do. Is it morally right to cheat? To answer this question, let's first consider a scenario.

On a dark and stormy night, you are chatting with your friend in front of the fireplace in your cozy home. Your TV is airing a warning about a fugitive on a killing spree when the doorbell rings. You go to check who the visitor is and can see a man through the door glass who you immediately recognize as the fugitive just shown on TV. He asks whether your friend is inside the house. Will you lie to the murderer about the whereabouts of your friend?

The story is a modern retelling of a dilemma posed by the eighteenth-century philosopher Immanuel Kant in his essay *On a Supposed Right to Lie from Altruistic Motives*.[26] Although most people choose to lie, Kant's answer is that you should not. Why? He names three reasons summarized by philosopher Jacob Weinrib: a lie "corrupts the source of right," "violates the duty of truthfulness," and "undermines the validity of contracts."[27] In Kant's own words, "Truthfulness is a duty which must be regarded as the ground of all duties based on contract." It is a "sacred and absolutely commanding decree of reason, limited by no expediency." That is, telling truth is an "intrinsic duty" that warrants no exception. The reason we should speak truth and only truth is because, Kant writes, "a lie always harms another; if not some other particular man, still it harms mankind generally, for it vitiates the source of law itself."[28] In other words, based on his categorical imperative, morality is not obeying the law; it is heeding your duty.[29] Telling the truth, therefore, is more important than protecting your friend's life.[30]

Kant's reason is repulsive in practice, however. It is the worst kind in the spectrum of the perils of honesty. Who would let a friend be murdered in order to stick to truth, whatever you think it is? What's wrong with this great philosopher, who was famous for his rigor in thinking? The answer: he mixed the legal and the ethical perspectives.[31] At least, he failed to make a clear distinction between the two domains, the descriptive versus the normative. How we behave in a court of law can be quite different from what we do in our daily lives.

What Kant had not resolved was answered by his contemporary Benjamin Constant. Constant noticed that rights and duties are correlated. Because "no one has a right to truth that harms others," then a killer has no right to the truth, and you are not bound by duty to be truthful to him. He writes,

> It is a duty to tell the truth. . . . The concept of duty is inseparable from the concept of right. A duty is that on the part of one being which corresponds to the rights of another. Where there are no rights, there are no duties. To tell the truth is therefore a duty, but only to one who has a right to the truth. But no one has a right to the truth, who injures others.[32]

Although the correlation thesis is obsolete today, both Constant and Kant agree that lying to the killer is not lying.[33] This approach, rather than solving the dilemma, introduces more problems. If lying to a killer is not a lie, what about lying to a friend who later becomes a killer? Is it a lie? How about lying to a person who is likely to use your information to spread a malicious rumor? Is it a lie?

These are just a few examples of problems we can expect to run into if we use Kant's advocacy for absolute honesty as a moral guide in our everyday lives. His position that no lie is permissible is not only too stringent to be practical but also gives rise to unresolvable dilemmas in application. Also, making the exception of lying to a murderer as "not lying" violates the objective definition of lying. That is, you can't declare that a lie is not a lie when it communicates false information ipso facto. Furthermore, if you make exceptions for some lies, you'll have to make a lot more exceptions for lies of a similar nature as well, a task that is as confusing as it is impractical. We would become mired in endless debates as to what constitutes an exception.

✿

Defining the boundary between permissible and impermissible lies has been a challenging philosophical quest since time immemorial. The most distinct strength in Kant's philosophical thinking is his axiomatic approach akin to formal mathematics. He constructed his ethics theory based on a set of self-evident truths, namely duties, for which his moral philosophy is known as deontological ethics. (Deontology is a word derived from Greek for duty or obligation.)

The logical strength of Kant's ethics is also its major weakness. This is because emotions are essential for us to make decisions and take actions, as recent research has shown. Emotions often override our long-held moral principles, especially when there is little legal consequence. Therefore, emotion-detached deontological ethics, despite its logical rigor suited for academics and court, has had limited impact on moral teachings.

Furthermore, not all people agree that telling the truth is a duty in their social lives. Rather than strictly sticking to a set of personal principles, most people simply follow social customs and conventions without questioning the legitimacy of these norms. As societies often differ in determining whether a certain lie is permissible, it is difficult for deontological ethics to come up with a set of universal criteria suited for all cultures that define which lies are permissible and which are not. We need to find an alternative framework in ethical philosophy to guide us in our everyday lives.

One method for deciding whether a lie is permissible is by its consequence in the real world, a moral philosophy formally known as, unsurprisingly, consequentialism. This is a position backed by utilitarianism,[34] a philosophy famous for its ethical advocacy of the greatest good for the greatest number of people. Although promoted by philosophers such as Jeremy Bentham and John Stuart Mill, the roots of consequentialism can be traced to ancient Greece. In the fifth century, for example, the Roman religious philosopher St. Augustine—although not identified as a consequentialist—analyzed the issue of cheating from a clear consequentialist standpoint, nonetheless. He put lies into eight categories, from the gravest to the least serious, indicating increasing permissibility:

Lies in religious teaching,
Lies that harm others and help no one,
Lies that harm others and help someone,
Lies told for the pleasure of lying,
Lies told to "please others in smooth discourse,"
Lies that harm no one and that help someone materially,
Lies that harm no one and that help someone spiritually, and
Lies that harm no one and that protect someone from "bodily defilement"[35]

What makes the consequentialist approach particularly appealing is that all normal humans share basic emotions and feelings, which can goad us to decide whether to embrace or reject an act (such as lying) based on its consequence. It is at this instinctive and intuitive level that all humans share a common moral ground as to what is good and what is bad on most issues, regardless of cultural differences.[36] Therefore, the consequentialist approach to cheating is most likely to reach a universal consensus.

Although Augustine's list is nonsystematic and too long to be practical, it inspires us to find a more complete and simple classification system for what is permissible and what is not. Here I entertain a proposal by sorting out all lies and deceptions into three categories:

1. *Prosocial cheating*: other-concerning, intended to respect, assist, or protect other's well-being
2. *Antisocial cheating*: selfish, intended to promote self-interest at the expense of others
3. *Self-serving cheating*: self-concerning but without perceivable harm to others

Based on this system, the answer to the question of what we *ought to* do becomes clearer. *Prosocial lies and deceptions should be permissible, if and only if no harm is done and there is no alternative for telling the truth.* In Kant's dilemma, lying to a murderer or an enemy is permissible—desirable, even—because it is prosocial and has no alternative. The same can be said for Oskar Schindler, who cheated and bribed the Nazis to save innocent Jewish people. This also gives a green light to most white lies in our social lives, and for physicians to use placebos when people's autonomy is respected. It also legitimizes what I did to save the ducklings by not telling my assistant Lao where they were hiding, in chapter 3.

The prosocial condition is critical to justify deceptive manipulation for social preferences. Therefore, the following cognitive manipulations should be permitted: arranging food in a cafeteria to nudge people to eat more healthily or resorting to a lottery system as a reward to encourage people to save more for the future. Similarly, during the campaign for Obamacare, the cost of the program was toned down to emphasize the overwhelming benefit of the program especially for low-income people.[37]

Note that this practical rule combines both a consequentialist and deontological approach. Besides consequence, it stipulates that a lie is allowed only when there is no alternative. As such, you shouldn't get a kick out of lying by claiming you went fishing yesterday when in reality you went for a hike. Likewise, prosocial lies that do not meet all conditions are not permissible. Consider the case of Casey Smitherman, the superintendent of a school district in Indiana. In January 2019, she used her son's insurance to get a student treated for strep throat in a clinic. Although her act was prosocial, she caused a loss for her insurance company. She also had alternatives not to lie, such as contacting the student's parents.[38] Likewise, a compliment is other-concerning, making it a prosocial act and giving it a better choice than withholding it. Conversely, flattery is self-serving and harmful to others, rendering it an antisocial act and therefore impermissible.

What if, you may ask, others do not agree with what we think is good for them? This is a potential caveat, difficult to resolve between different peoples or cultures. For instance, lying about drinking coffee is harmless in most situations, but for Mormons, it is doubly offensive. Similarly, American physicians have an unnegotiable duty to tell a patient the truth regarding the diagnosis of a fatal disease such as a terminal-stage cancer. Lying about the condition of the patient would breach the principle of informed consent between physician and patient. For Chinese physicians, on the contrary, lying to a patient is considered prosocial because keeping patients in the dark can prevent them from being devastated by the sudden bad news.[39] Therefore, we should keep an open mind, judging the consequence of a lie according to the norms practiced in a specific community or culture.[40]

Antisocial lies and deceptions are impermissible under any situation. This type of cheating is what we mostly refer to when we say, "All lies are bad." Thus, shady practices by a political party to win elections are not permissible, because they hurt at least one group of voters—if not both—in a bipartisan system like that in the United States. In the same vein, using social networks and TV programs to *willfully* spread false information—*disinformation*—is unethical because it can hurt innocent people at the receiving end.

The most difficult distinctions come with self-serving cheats such as most boasts and advertisements. As philosophers such as Aristotle and Kant realized, all lies breach social trust. Thus, self-serving cheats should never be encouraged. However, if cheats cause no perceivable harm to others, they should be tolerated to a certain degree, depending on the prevailing norms for a particular community or the larger society. Commercial advertisements, for example, should be allowed as long as their level of deception falls within the norm practiced in the industry and accepted by society at large. Still, conventions may not always make this type of cheating permissible. For instance, the lies and deceptions used to convince people to buy real estate during the boom of the early 2000s should have been impermissible. With hindsight, we can see those lies fall into category 2 in the proposed classification system. However, that information wasn't available to the public at the time.

𝟊

Information is one of the stickiest issues. Falsifying it or bending it, even hiding it for a specific purpose—a lie of omission—is a form of deception. Can we live a moral life without causing harm or being harmed by others? This again brings up Kant's idea of the right to know the truth. The question is: Do all people have an equal right concerning information?

Kant, arguably the greatest philosopher since Aristotle, was born in Königsberg in East Prussia (now Kaliningrad, Russia, on the Baltic Sea) and spent his entire life in the city. He lived a stoically disciplined life without venturing out more than ten miles, except between 1750 and 1754, when he worked as a private tutor. He confined his activity to his small, quiet neighborhood, where he knew virtually all of the residents.[41] Legend has it that he took an afternoon walk every day along a fixed path at four o'clock—he was so precise that his neighbors trusted him more than the clock on the town cathedral. Since he never had to worry about falling prey to scams, he had the luxury of sharing information freely with all the honest and trustworthy folks in his small, close-knit community.

Kant's default position of "right to know" doesn't extend to all situations, however. He realizes it becomes problematic when dealing with criminals or enemies, and his solution is to revoke the right. But that poses a new conundrum: Does a person have the right to know the truth from you *before you know who they are*? In my own experience, I have had several research ideas stolen from my lab during free exchanges at academic conferences. How could I know that colleagues from other labs would plagiarize our ideas? Who has the right to know is even more of an issue in the digital space, where most of the time, we have little knowledge about who we are dealing with in commerce, scientific communication, consulting, and many other social interactions.

For most of us today, society is far larger than Kant's Königsberg neighborhood. Frequent direct dealings with complete strangers aside, the Internet puts us in contact with a community that includes the entire world. Most people we interact with in this digital universe are faceless or even nameless agents we will never meet face-to-face and are from faraway places we have never been and will never travel to. If we were to continue offering information as freely as Kant apparently did to those in his immediate neighborhood, we make ourselves vulnerable targets for digital scams. For this reason, protecting our private information by whatever means necessary has become a top priority in the modern world.

Although it's hardly a desirable approach, many of us use false birthdays, false home addresses, even false names on social media, such as Facebook, LinkedIn, and Twitter. When we resort to this defensive measure, we're already lying because we've falsified the truth. Although safeguarding your private information is by its nature a dishonest act, it can protect you from being hurt. As such, it is self-serving without causing harm to other innocent people. This is permissible because it falls into category 3 in our classification system.

In the same sense, it is morally justifiable to keep a secret for others. Although it is a form of dishonesty, considered by many a burden to bear,[42] failing to do so is unacceptable. It is even illegal for some professionals such as lawyers, physicians, bankers, and psychologists. In a sense, we are turning Kant's position of right to know upside down as our default position for private information, particularly in the Internet Age. That is, *people have no right to know the truth unless they are granted the right*. So, when a friend of yours confides in you and asks you to keep a secret for her, she is granting you the right. Even so, the right is not transferable to others without her permission. Thus, being a big mouth or snitch is unethical. This position agrees with our right to privacy according to Article 12 in the 1948 Universal Declaration of Human Rights.

Having said this, the right to know the truth remains the default position for *public information*. It is intrinsic to the nature of a democratic society that citizens are automatically given the right. In principle, a democratic government is not permitted to lie to its citizens. Even classified information, which should not be disclosed immediately, should still be subject to the principle of transparency when it is safe to reveal it to the public at large.

𝒜

For Christians, humans are born with sins; for Confucians, humans are born good; for Buddhists, humans are born to suffer. As the true nature of the human species is increasingly understood by science, such sweeping generalizations are far too simplistic. We are more complex, capable of being both good and bad. As a scientist, I am mostly an optimist about human nature, a view that is borne out by research, among which is a recent real-life experiment led by social scientist Alain Cohn.[43]

In the study, researchers placed items inside a wallet: three identical business cards (which provided contact information of the wallet's "owner"), a grocery list, and a key. Sometimes they added a small amount of money ($13.45 in local currencies), and sometimes they did not. They then gave the wallet to a random person, claiming it was lost by somebody in the street. They repeated the procedure in 355 major cities in 40 countries. Although the rate of wallet return varies, one trend is overwhelmingly consistent: a lost wallet is more likely to be returned when it has money inside it than when it has no cash. This demonstrates that people generally have concerns for others' welfare. It's a testimony to the evolutionary wisdom inscribed in human nature: win-win cooperation wins out over win-lose manipulation. That's why antisocial cheating is nearly always a minority strategy, both in humans and in other animals.

Even so, antisocial cheating can still cause major problems. In the United States, tax cheats alone can lead to an annual loss of several hundred billion dollars in government coffers. In developing nations, corruption and illicit financial dealings can take an even heftier economic toll, amounting to 1.3 trillion dollars a year.[44] Dirty politics, tax evasion, and corruption can further discourage honesty, leading to the loss of social trust.[45] This in turn makes people pay higher costs in economic transactions alone, not to mention the divisive social effects of living in a low-trust milieu. Clearly, fighting against antisocial cheating by any means is well worth the cost.

Although it's a long-standing problem in all human societies throughout history, antisocial cheating poses a slew of new challenges today, not simply because it continues to exist, but because it has spread from the physical world to the digital universe, with ever-increasing reach, speed, and impact. Considering the prevalence of spam emails (29 billion sent every day), a single scam may easily and quickly affect millions of people across the world. As an FBI report shows, the financial losses to Internet-based theft, fraud, and other schemes amount to $2.7 billion a year. From phishing tricks to lonely-heart scams, our cognitive biases and psychological weaknesses are being exploited to the limit.[46]

What's more, democratization of information online has spurred the creation and spread of false information, making us all potential victims and victimizers, with or without nefarious intent. As false information and fake news displace accurate and reliable facts, our ability to make good decisions is compromised at both personal and societal levels. Take elections, for example. Since hackers can generate outsized influence by manipulating information through social media, the rules and norms of democratic processes have been dethroned, no longer serving as guardians of the majority's will.[47] The consequences can be grave.

In 2018, the Policy Planning Staff, a think tank within the French Ministry of Foreign Affairs, sent a dire warning about information manipulation that exploits our cognitive loopholes, especially confirmation bias, as a major threat to democracy worldwide.[48] Just slightly over two years later, the warning became real in the United States. On January 6, 2021, thousands of violent rioters stormed the Capitol Building and attacked Congress, while chanting "Hang Mike Pence!" The rioters vandalized the building and sent members of Congress into hiding to save their lives. In more than two centuries, American democracy had never been so seriously challenged.[49] And all of it started on the false claim that the 2020 election was stolen from Trump. Who had expected that disinformation could suddenly become a major challenge to the survival of American democracy? Fighting disinformation, consequently, has become a new and urgent duty for citizens.

If history has anything to tell us, it's in its wrongheaded approach to fighting against cheating. For much too long, we've attempted to stamp cheating out by any means, including laws, mores, religious teachings, and philosophical theories. For instance, four of the Ten Commandments in the Bible deal with lying (adultery, stealing, bearing false witness, and coveting). One of Buddhism's Five Precepts is speaking for truth. In both the East and the West, we are told from a very young age that we shouldn't lie under any circumstance. We are told we'll be penalized by lightning strikes or growing a long nose. Still, can you imagine a time when there were no scammers, hustlers, swindlers, and con artists? How many of us really believe that cheating can be thoroughly banished from society? The lofty goal of completely eradicating cheating has in reality trapped us in a dead end. It's truly astounding to know we've been on the wrong track against cheating throughout history!

This no-win situation in getting rid of antisocial cheating doesn't mean we're doomed, but that we must rethink our approach. Rather than put our time and effort into the impossible mission of eliminating cheating entirely, perhaps we should instead turn to a more feasible goal.

Many experts have provided practical advice for fighting scams and frauds. For example, law professor Tamar Frankel suggests several useful ways to identify red flags (mostly, "too good to be true") and defend against fraud in her book *The Ponzi Scheme Puzzle*. She also advises separating emotion and faith from business to protect our cognitive loopholes from exploitation.[50]

Although useful, these case-based, area-specific methods for fighting fraud are far from sufficient because the online environment today exposes all of us to a wide range of novel scams, not just one type, as was mostly the case in the past. We need a generic tool, like a Swiss Army knife, to deal with all kinds of false information, whether they are spam, scams, or fake news.

One emerging idea to develop such a tool looks at how our immune system handles infectious diseases and is based on fighting misinformation/disinformation through inoculation—that is, by exposing ourselves to a weakened version of it, like a vaccine, we can strengthen our cognitive resistance. That way, when a real scam or a weaponized version of fake news strikes, we will have already gained some degree of immunity against its manipulative effects.[51]

This immune system–inspired idea is potentially fruitful and practical for two reasons. First, pathogens and scams are not just superficially similar. They share fundamental features when interacting with people because both take the form of an evolutionary arms race. In theory, there is no difference between the battle of mutant HIV strains against anti-HIV treatments and the battle of computer viruses against antiviral programs. Our scientific knowledge garnered from fighting germs should in principle work for fighting against digital fraud as well. Second, the COVID-19 pandemic, although a global disaster, has provided

TABLE 8.1. Germ-Inspired Principles for Fighting Antisocial Cheating

	Disease Germs	Antisocial Cheating
Origin	Mutation, transmission	Invention, modification, transmission
Attack point	Weaknesses and deficits in the immune system	Biases, weaknesses, and deficits in the cognitive system
Harm	Physical health	Physical health, financial health, social health, political health, etc.
Preventive measures	1. Avoid potential sources 2. Sanitize 3. Vaccinate	1. Avoid potential scams 2. Educate 3. Warn
Cure	Antibiotics	Law enforcement
Campaign objective	Contain (e.g., herd immunity)	Contain (e.g., societal immunity)

an opportunity for vast numbers of people to understand how our immune system works. This helps broaden our knowledge from fighting infectious disease to fighting online fraud and information manipulation of all kinds.

Drawing from the evolutionary wisdom about cheating revealed in this book, I put forth a proposed blueprint, summarized in table 8.1, to fight antisocial lies and deceptions.

The proposed strategic blueprint can be effectively adapted for practical use when dealing with a specific type of antisocial cheating, whether scams, fraud, or any variety of information manipulation. Figure 8.1 is an example of a public campaign against Internet and cell phone scams in the city of Hangzhou in China.

Besides preventive means, as the Chinese example illustrates, we can also resort to policing for fraud control. Facebook and Twitter, for instance, currently ban people from spreading disinformation. Even the then-president Trump was barred from using the social media for sending false information and relaying conspiracy theories, especially about the results of the 2020 presidential election.

FIGURE 8.1. Public campaign against digital frauds, displayed on a bus in the city of Hangzhou, China in 2019 (photo credit: Lixing Sun).

Such "hard policing" has two obvious faults. First, it's reactive, mostly serving as damage control after the damage is done. Its effect is quite limited after the false information has already become extensive in social media, just as quarantining a few patients helps little in controlling a contagious disease after the germ is already widespread. Second, hard policing can also raise the question whether private companies can and should revoke the right of free expression, especially when users try to voice their political views. These problems goad us to seek proactive "soft policing" methods, which are both more acceptable and more effective in reducing disinformation in society. Do such ideal solutions exist?

The answer is yes. Here is one such new idea. YouTube, currently the world's most visited website, is also a primary source of disinformation, thanks partly to some YouTubers. Although YouTube, like Facebook

and Twitter, resorts to strict policing to control disinformation, its effect
is limited, as expected. A main problem lies in a business practice called
the YouTube Partner Program, where YouTubers receive ad revenue of
$3–5 per thousand views. So, if you post a video clip that can reach one
million views, you can pocket $3,000–5,000. This is how top YouTubers
can make as much as $20 million a year. It's a no-brainer then that most
YouTubers prioritize building their fan base. And some resort to cheating by exaggerating, radicalizing, and fabricating information or stories.
That way, they can draw more people to watch their videos.

An easy fix, therefore, is to reward YouTubers who tell the truth. To
do so, YouTube can hire fact checkers or use a program to periodically
rate the truthfulness of video information and tie it to pay rate. This will
reward honest YouTubers and punish spinners of false information and
conspiracy theories. Such a truth incentive, which can be readily implemented, is soft, proactive, and effective in cutting down disinformation.
(Please help me persuade YouTube to do so!)

𝕏

Despite its negative effects, antisocial cheating can serve as a catalyst for
innovation, both biologically and culturally. Cybersecurity businesses,
for example, wouldn't sprout, grow, and bloom without the presence of
malicious cheating. Among the many beneficiaries is CrowdStrike, a
cloud-based cybersecurity company initially headquartered in Sunny-
vale, California.

CrowdStrike went public on June 12, 2019. Just one day earlier, it was
priced at $28–30 per share. But the demand from investors was so high
that its price had to be raised to $34 a share just hours before it was open
for trading. By the end of the first day as a public company, Crowd-
Strike's stock price was $50.10, more than 70% above its IPO price. And
the company was worth $11.4 billion.[52]

Since its founding in 2011, CrowdStrike has achieved several land-
mark accomplishments. Among them was the identification of Russia
as the source of the hack into the DNC computer in the 2016 US presidential election. This might have given Donald Trump just enough help

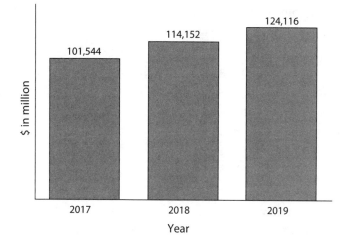

FIGURE 8.2. Worldwide spending on cybersecurity in recent years. Data from Gartner research and consulting firm; see Roger Aitken, "Global Information Security Spending to Exceed $124B in 2019," *Forbes*, August 19, 2018.

to win the electoral college vote, which has vastly altered the American political landscape.

Dimitri Alperovitch, cofounder of CrowdStrike, was rumored to have made this boast: "There are two kinds of companies: those that have been hacked and know it; and those that have been hacked and don't." And "those that buy CrowdStrike," chimed in Richard Clarke and Robert Knake in their book *The Fifth Domain*. According to the authors, the digital economy is growing two times faster than the traditional economy. Yet cyberattacks extract a high cost. For instance, malware by the name of NotPetya led to billions of dollars in losses for companies doing business in the US and Europe, particularly, Ukraine.

Understandably, worldwide spending on cybersecurity has skyrocketed—$114 billion in 2018 alone (fig. 8.2). In the US, $15 billion in the president's budget was dedicated to cybersecurity. By 2021, cybercrime damages reached $6 trillion per year across the planet.[53] As a result, cybersecurity companies have experienced exponential growth, topping 3,000 in number in recent years. The cyber insurance business has increased quickly as well.[54] Without out-of-control cheating in the digital space, companies like CrowdStrike would not even exist, much less reach such high valuations so fast.

Although malicious lies and deceptions—especially those that are Internet-based—are impossible to eliminate, they spur technological innovations aimed at containing them. And unlike fighting cheating schemes before the Internet Age, we all have a vital stake in the new arms race we're fighting now.

✺

In this book, we have explored in depth the two laws all cheaters use: alter truthful information in communication (the biological essence of lying) and exploit cognitive loopholes (the biological foundation of deception). Those laws apply equally in the biological world as well as in our social and cultural realms. Understanding these fundamental principles can help us design effective ways to fight antisocial cheating. This is quite like girding ourselves against destructive biological agents: germs, diseases, and pests. Both are examples of evolutionary arms races. Instead of attempting to eliminate them from the face of the earth, a goal that has been proven impossible time and again, it's more feasible to contain them.

As we've seen, one of the most surprising ideas in this book is that cheating—contrary to popular belief—is a powerful catalyst in creating diversity, complexity, and even beauty in nature. Cheating has led to novel behavioral tactics, physiological adaptions, and morphological structures. It has paved the way for the emergence and sophistication of intelligence and art. However, cheating has been so profoundly maligned in our culture that we tend to forget its important role in driving biological and cultural diversity.

If, before you read this book, you did hold a pessimistic view of cheating, based on the prevalence of malicious lies and deceptive tricks in our society, I hope you feel somewhat relieved at this point. In this regard, we should appreciate the German philosopher Hegel's take: "What is rational is real; and what is real is rational."[55] Certainly we shouldn't embrace antisocial cheating as good, but we can make the best of it. By containing it, we will not only lessen its harmful impacts but also harvest its catalytic power for innovation and the advancement of science, technology, economics, education, law, and many other aspects of our culture. Cheating and anti-cheating will continue to be odd bedfellows whose arms race will go on to stimulate new, positive developments. Therefore, we should not only take the Daoist stance of accepting cheating without fear or despair, but also have the audacity to rise above it and thrive amid its existence.

ACKNOWLEDGMENTS

The idea of this book was conceived a decade ago when I completed the *Fairness Instinct* manuscript. At the time, I wanted to keep my writing momentum going. Yet, I also needed a diversion from focusing for too long on a taxing and serious topic. So, I thought a book about cheating would be fun to work on. A few years into this project, I realized that I was wrong—dead wrong—especially after the COVID-19 pandemic started. Luckily, effective vaccines were quickly developed. I thought they would immediately free us from the onslaught of the pandemic. Unfortunately, people continued to suffer, thanks to vaccine disinformation, which ravages us like a toxic gas pervading our media and digital space.

I used to see politicians, public figures, and media gurus as modest, decent, and respectful. Today, so many of them lie brazenly, making false claims and promoting conspiracy theories, as if fact and truth no longer matter. What matters are ratings and fan bases for popularity, which ultimately boils down to power and money. One major ramification of a post-truth society is political extremism. Before the fateful day of January 6, 2021, I never thought American democracy could become vulnerable when challenged by disinformation.

These developments prompted me to venture into several new aspects of cheating that were not originally planned, making the book far broader and more difficult to write. Luckily, I've been surrounded by a great team who made the completion of the book possible. Although my name is on the byline, this book is the fruit of a collective effort.

First in line is my editor, Alison Kalett, who spotted the potential of the book and guided me through the entire writing process from the beginning to the end. In addition to providing editorial insights, she was a tireless spokesperson for the book. Without her, you would not see this book today. Yet, Alison is not the only one at PUP who stood behind the book. Hallie Schaeffer was another major force throughout the incubation process of the book. She was always available for solving issues large or small. My gratitude also goes to Lisa Black, whose expertise made an otherwise continuing struggle in obtaining copyrights a painless process. I was really lucky to have Susan Matheson as my copyeditor, who, in addition to ironing out rough spots in the manuscript, caught and killed well-camouflaged errors that had escaped my eyes. My thanks also go to Jill Harris, David Campbell, and many others at PUP for help and encouragement during the production of the book.

Writing a science book that is accessible to an intelligent audience, both with and without subject knowledge, is never easy, especially in a language radically different from one's mother tongue. This difficulty would have been insurmountable had there been no linguistically superior people who were keen to offer their help. For this, I thank a stream of graduate and undergraduate students from Central Washington University's writing center under the directorship of Jared Odd. Dave Darda also gave me many good comments on earlier versions of the first three chapters. Matt Altman was extremely instrumental and insightful in philosophical issues and provided critical help with the discussion about the moral concerns of cheating. I also received strong support from my friend Alan Honick, who edited line by line the entire manuscript before it was finalized. My most available editor, however, was in-house literally: my son Orien. Orien not only edited all versions of all the chapters in every possible way—wording, sentence structure, grammar, ethos, logic, science—but also became an inexhaustible source of fun and encouragement, which gave me comfort and kept me writing over the years. My older son, Shine, known for his golf ball accident in chapter 4, is now known in the Sun family for his broad knowledge and ability to think deeply. It was he who came up with the solution for protecting private information in the digital age, discussed in chapter 8, by turning Kant's moral position of "the right to know the truth" upside down.

Three anonymous and highly competent biologists also provided their insightful comments throughout the book. Their scientific knowledge and expertise helped uphold the rigor of the science presented in the book.

The book would feel far less vivid were there no images along with the stories. In this regard, many individuals generously and enthusiastically offered their help by sending me or allowing me to use their images. Here, my heartfelt gratitude goes to Andrew Bass, Jianguo Cui, Keven Drew, Szabocs Kokayart, Jingang Li, Dingzhen Liu, Dietland Müller-Schwarze, David Nash, Elizabeth Peters, Gil Rosenthal, Fuwen Wei, Rongping Wei, Chaocan Yang, Yue Yang, and Janxu Zhang. I also thank those who have made their beautiful images freely available in the image collection of the Creative Commons project. Their generous donations have made this book less costly and more accessible to readers.

Finally, I thank all my immediate family members—Crystal, Shine, and Orien—for their enthusiastic support by keeping me happy and freeing me from family chores over the years.

NOTES

Chapter 1. Liar, Liar, Everywhere

1. Ghoul, Griffin, and West (2014).

2. For some biologists, deception is considered cheating only when an organism manipulates a cooperative system, but it's not viewed as cheating when an organism adapts to environmental conditions such as using mimicry. This subtle distinction is necessary for the accuracy in research communication.

3. Jersakova, Johnson, and Kindlmann (2006).

4. Müller-Schwarze (2006).

5. Truffles are highly valued—to a degree that borders on the ridiculous. European white truffles can sell for as much as $4,000 a pound. The current record is $330,000 (2007) for a 3.3-pound (1.5 kg) white truffle from Italy. The reason why truffles are so prized may be because the fungal steroid is also found in humans, especially in sweaty armpits. Sniffing the steroid, it is believed, may rouse your sexual instinct, making your mate seem more attractive to you. Apparently, sex sells.

6. Strassmann, Zhu, and Queller (2000).

7. Khare and Shaulsky (2010).

8. Santorelli et al. (2008).

9. Griffin, West, and Buckling (2004). Butaitė et al. (2016). Bruce et al. (2017).

10. Bruce et al. (2017).

11. Turner (2005). Díaz-Muñoz, Sanjuán, and West (2017).

12. The topic is hotly debated. Apparently, some junk DNA may have yet-to-be-known functions, but much of it may indeed be junk that has gotten a hitchhike in our genome over time. See Palazzo and Gregory (2014).

13. Hurst and Werren (2001).

14. Batzer and Deininger (2002).

15. Sun et al. (2012).

16. Hancks and Kazazian (2016).

17. Burt and Trivers (2006). Bourke (2011).

18. Rice (2013).

19. Hurst and Werren (2001).

20. Kiers et al. (2003).

21. Porter and Simms (2014).

22. Sosis and Alcorta (2003).

23. Lixing Sun, "Would Twitter Ruin Bee Democracy?," *Nautilus*, December 14, 2017, http://nautil.us/issue/55/trust/would-twitter-ruin-bee-democracy.

Chapter 2. Hackers and Suckers in Communication

1. Even though two individuals are rivals, they can still benefit from communication, where the rule of engagement will be followed, and unnecessary cost (such as time, energy, and injury) can be avoided for both.

2. Dawkins and Krebs (1978).

3. Here is the calculation: $1 \times 34.21\%$ (if it's a scam) $+ 3 \times [1 - (34.21\%)]$ if it's a real deal.

4. Ghoul, Griffin, and West (2014).

5. Zuk, Rotenberry, and Tinghitella (2006).

6. Maynard-Smith and Harper (2003). Scott-Phillips et al. (2012).

7. Bugnyar and Kotrschal (2004). Bugnyar and Kotrschal (2002).

8. Steele et al. (2008).

9. Coussi-Korbel (1994). Hauser (1992). Anderson et al. (2001).

10. Hirata and Matsuzawa (2001). De Waal (1982). Goodall (1986).

11. Moller (1990).

12. Tamura (1995).

13. Plath et al. (2008).

14. De Waal (2019).

15. Zhang, Sun, and Novotny (2007).

16. At that time, I was too little and ignorant to realize that China claims a thousand years of history in cricket fighting as a hobby and a blood sport. Today, many cities hold cricket-fighting tournaments, often with meticulous classification of weight classes, like boxing, for fair competition. Beijing even hosts annual national championships for cricket fighting. Unsurprisingly, live-cricket trading has grown to be a multimillion-dollar business today.

17. Steger and Caldwell (1983).

18. McLain et al. (2010).

19. Bee, Perrill, and Owen (2000). Sullivan-Beckers and Crocroft (2010).

20. McQuire et al. (2018).

21. Byrne and Whiten (1985). Byrne and Whiten (1990).

22. Slocombe and Zuberbühler (2007).

23. Baglione et al. (2010).

24. Fan et al. (2018).

25. Heinsohn and Packer. (1995).

26. This unique mechanism is called haplodiploid sex determination and is common in bees, wasps, and ants. In these social insects, unfertilized eggs with only one set of the genome (1n, or haploids) develop into males, whereas fertilized eggs with two sets of the genome (2n, or diploids) develop into females.

27. Nonacs (2006). Beekman and Oldroyd (2008)

28. Riehl and Frederickson (2016).

29. Nunn and Lewis (2001).

30. Wickler (1968).

31. Smoltification transforms the small, stream-bound rainbow trout into large, seafaring steelhead salmon. The two drastically different forms made many believe for some time that they were two separate species.

32. Bass (1996).

33. Fergus and Bass (2013).

34. Sinervo and Lively (1996).

35. Whiting, Webb, and Keogh (2009).

36. Mason et al. (1989).

37. Crews and Garstika (1982).

38. Mason et al. (1989).

39. Mank and Avise (2006).

40. Gerhardt and Huber (2002).

41. Such information asymmetry is common in card games where players know their own hands fully but must guess the cards in others. Therefore, bluff and bravado are part of the game in card tournaments such as Texas Hold'em, quite unlike chess matches, where information is all there in the open, symmetrical for both players. In games where information is symmetrical for players, tricks and cheats are harder to contrive and are often ineffective between equals, for which explicit bluff and bravado are less commonly used.

42. Hare and Atkins (2001). Pollard and Blumstein (2012).

43. Hypersensitivity is adaptive for young animals. It's better for them to overreact than underreact before they learn to know which signs are dangerous and which signs are not.

44. Riehl and Frederickson (2016).

Chapter 3. Nature's Eavesdroppers, Impostors, and Con Artists

1. Stevens (2016).

2. If you are familiar with the works of Salvador Dali, his whale-sized space elephants can only live in the artist's surreal world.

3. Interestingly, people who are blind can be trained to use sound to map objects around them by echolocation, and some can be quite good at it. The information-processing center appears to lie in the primary visual cortex. See Norman and Thaler (2019).

4. The term "sensory bias" has been recently broadened to perceptual bias; see Ryan and Cummings (2013). I think replacing the word "perceptual" with the more general "cognitive" may be more inclusive. Here I keep the term "sensory exploitation" because it is already well-known in the literature.

5. Schaefer and Ruxton (2009).

6. Eberhard (1977).

7. Stegen, Gienger, and Sun (2004).

8. Technically, short-term color change is termed "physiological color change," whereas long-term color change is termed "morphological color change." Although it's not entirely clear yet as to the difference between the two types of color change, they both involve pigment cells on the skin known as chromatophores.

9. Wallace (1867).

10. Stevens (2016).

11. Vallin et al. (2005).

12. Vallin et al. (2005). Vallin, Jakobsson, and Wiklund (2007).

13. Caro et al. (2014). Caro (2016).

14. Kojima et al. (2019).

15. Stevens (2016).

16. Stevens (2016).

17. Darwin (1859).

18. This can happen in human societies after a major war, such as in Europe after World War II or in Korea after the Korean War. In both cases, the number of surviving men was so much less than the number of women that a de facto polygyny was officially or unofficially sanctioned in places.

19. Trivers (2011).

20. Hanlon, Forsythe, and Joneschild (1999). Hanlon et al. (2005).

21. Müller (1879).

22. Batesian and Müllerian mimicries are generally recognized among biologists as two fundamentally different types of mimicry. Yet, they converge in terms of how they work: both involve the exploitation of the cognitive systems of predators.

23. Stevens (2016).

24. Cheney (2012).

25. Hafernik and Saul-Gershenz (2000).

26. Saul-Gershenz and Millar (2006).

27. Kikuchi and Pfennig (2010).

28. Stevens (2016).

29. Kelley et al. (2008).

30. Barber and Conner (2007).

31. Nelson (2012).

32. Nelson and Jackson (2009).

33. Als et al. (2004).

34. Barbero et al. (2009).

35. Akino et al. (1999).

36. Stevens (2016).

37. Allies, Bourke, and Franks (1986).

38. Stevens (2016).

39. Gilbert (1982).

40. Kurup et al. (2013).

41. Bauer et al. (2015).

Chapter 4. Infidelity and the Rise of Honesty

1. Shine was "bankrupt" after insisting on paying his half with his lifetime savings in his piggy bank—$77.

2. Ratnieks and Wenseleers (2005).

3. Dugatkin (1992).

4. Dugatkin (1991).

5. Dugatkin (1997).

6. Eukaryotic organisms refer to those whose cells contain nuclei and organelles. They include such organisms as protists, algae, fungi, plants, and animals. In contrast, prokaryotic organisms refer to those whose cells do not have nuclei and organelles. They include bacteria and archaea. Also, bacteria have several ways to diversify their genetic makeup. The best known is conjugation: two bacteria connect themselves with a tiny tube through which circular DNA molecules, known as plasmids, can be swapped.

7. In terms of relative size, the kiwi egg is the largest. It can account for a quarter of the kiwi's body mass. In humans, about 400 eggs are produced in a woman's life. In contrast, the number of sperm cells in a single ejaculate can be several hundred million, and in a man's lifetime, about a half trillion.

8. "Most Prolific Mother Ever," Guinness World Records (website), https://www .guinnessworldrecords.com/world-records/most-prolific-mother-ever?fb_comment_id =84106.

9. Such classic stereotypes of males and females are often wrong for gross simplification of the biological reality, which is far more complex and nuanced. For instance, many species of female songbirds have recently been found to make calls, which are typically viewed as a male reproductive strategy. See Riebel et al. (2019).

10. Westneat (1987).

11. Griffith, Owens, and Thuman (2002).

12. Gerlach et al. (2012).

13. Arnqvist and Kirkpatrick (2005).

14. This is more formally known as the sexy-son hypothesis. Here is how it works. If you take a close look at Bateman's rule in polygynous species where a male mates with two or more females, you will know that having a sexy son is a genetic home run because it will increase your reproductive success by multitudes, something that even an extremely fertile daughter can't keep up with.

15. Gerlach et al. (2012).

16. In humans, fathers prefer children whose facial features resemble theirs. One study, by manipulating digital images of children, shows that men, more than women, are willing to invest more time and money on children who look like themselves than on those who don't. See Platek et al. 2002.

17. Kempenaers and Schlicht (2010).

18. Watts (1989). Sommer (1994).

19. Bruce (1959).

20. The sunk-cost fallacy refers to the tendency in humans to put in more money or effort when in a losing position, hoping to recoup the lost investment. It's related to our mental bias of risk aversion.

21. Because few people had trapped more beavers than I had, I proudly declared myself the reincarnation of a Mountain Man, a moniker for colonial American beaver trappers. By the way, beaver castoreum was prized as a main source of high-end perfume at the time.

22. Sun and Müller-Schwarze (1998). Sun (2003).

23. Zhang et al. (2007). Liu et al. (2008).

24. This is known as optimal breeding, by which animals can avoid both extreme inbreeding (such as mating with a close relative) and extreme outbreeding (such as mating with individuals of another species), both of which can lead to lower fitness.

25. Syrůčková et al. (2015).

26. Crawford et al. (2008).

27. My research records indicate that adult males in some of the complex families appeared to be brothers. So, their offspring could be cousins living in the same colony. These records, however, have not been genetically confirmed.

28. Cui, Tang, and Narin (2012).

29. Chen et al. (2019).

30. Evolutionary biologist R. A. Fisher proposed a hypothesis to account for evolution of the peacock's tail in 1930. But it appears too restricted to explain a broader pattern. We will talk about Fisher's idea in the next chapter.

31. Zahavi (1975).

32. Hamilton and Zuk (1982).

33. Zhang et al. (2008).

34. Zahavi and Zahavi (1997).

35. Barsh (2016).

36. Mundy et al. (2016). Lopes et al. (2016).

37. Hagelin (2002).

38. Tibbetts and Izzo (2010).

39. Bshary (2002).

40. Fitzgibbon and Fanshawe (1988).

41. Andrews et al. (2017).

42. That's why anti-doping is taken seriously in many professional sports today, for people want to see honest showing of capability, not cheating.

43. Strassmann (2003).

44. Greitemeyer, Kastenmüller, and Fischer (2013).

45. Bird and Smith (2005).

46. Bird and Smith (2005).

47. Veblen (1899).

48. Densley (2012).

49. Cloud and Taylor (2019).

50. Ian Steadman, "'Trillions of Carats' of Diamonds Found under Russian Asteroid Crater," *Wired*, September 18, 2012, https://www.wired.co.uk/article/russian-diamond -smorgasbord.

51. Barclay and Willer (2006).

52. Lyle, Smith, and Sullivan (2009).

53. Irons (2001).

54. Wood (2016).

55. Iannaccone (1994). Olson and Perl (2005).

56. Wilkinson (1990). Carter and Wilkinson (2013).

Chapter 5. Catalyst for Innovation

1. Readers can listen to the full song on this YouTube site: Sonya Spence, "No Charge," July 18, 2013, www.youtube.com/watch?v=N0f7K6CyZ14.

2. Because many birds can count the number of eggs in their nests, swapping eggs may avoid suspicion from the host.

3. Don't laugh, please. We also do similar things, such as putting small, flat images of our loved ones in our wallets or cell phones, thinking they are real.

4. Additionally, warblers also resort to other means (such as mobbing) to keep cuckoos away from their nesting areas; see Davies and Welbergen (2009).

5. In behavioral economics, this is known as base rate. Its calculation requires Bayesian statistics.

6. Davies, Brooks, and Kacelnik (1996).

7. Feeney, Welbergen, and Langmore (2014).

8. Trivers (2011).

9. Soler, Pérez-Contreras, De Neve (2014).

10. Honeyguides are African birds that often communicate with honey badgers or humans to indicate where honey is located. When the birds discover a bee nest, they emit a special call to guide their recruits to the nest.

11. Tanaka and Ueda. (2005).

12. Colombelli-Négrel et al. (2012).

13. Hoover and Robinson (2007).

14. Hoover and Robinson (2007).

15. Lyon and Eadie (2008).

16. In insects, evidence shows that egg dumping (sneaking eggs into the nests of members of one's own species) may not necessarily be detrimental to the hosts. The relationship, therefore, can be mutualistic rather than parasitic.

17. Michener (2000).

18. Cosmides and Tooby (1992). For a more generic introduction, see Christopher Badcock, "Making Sense of Wason," *Psychology Today*, May 5, 2012, www.psychologytoday.com/us/blog/the-imprinted-brain/201205/making-sense-wason.

19. Some researchers use "social selection" or "cultural selection" for this specific component of selection.

20. Naked mole rats also evolved a eusocial system due to close genetic relatedness.

21. DeCasien, Williams, and Higham (2017).

22. Ashton, Thornton, and Ridley (2018).

23. The relationship between brain size and group size is still controversial. Some studies support it, such as Street et al. (2017), whereas others refute it, such as Powell et al. (2017).

24. Dunbar and Schultz (2007).

25. Lindenfors, Nunn, and Barton (2007).

26. Byrne and Whiten (1992).

27. Byrne and Corp (2004).

28. Krupenye et al. (2016).

29. Gopnik (1993).

30. De Waal (2019).

31. Kiazad et al. (2010).

32. Bereczkei et al. (2015).

33. Wilson, Near, and Miller (1996).

34. Barrett and Henzi (2005).

35. Byrne (2018).

36. Bell and Buchner (2012).

37. Levine (2019).

38. Dunbar (1998).

39. Dunbar (1992).

40. Gonçalves, Nicola, Alessandro (2011). Norwitz (2009).

41. Dunbar (1998). Dunbar (2004).

42. Talwar and Crossman (2011).

43. The answer is one, Kareem Serageldin of Credit Suisse, compared with 25 bankers associated with banks in Iceland, 11 in Spain, and 7 in Ireland. One main reason why American bankers and financial executives avoided punishment may be because they had discovered and exploited loopholes in the American political and criminal justice systems.

44. Burley (1988).

45. Basolo (1990).

46. The methods they used were local squared-change parsimony and squared-change parsimony, a bit more sophisticated than averaging the calls of the two species.

47. Ryan and Rand (1999).

48. Rosenthal and Evens (1998).

49. Christy (1995). Proctor (1991).

50. Hughes et al. (2015). Fernandez and Morris (2007).

51. Burley and Symanski (1998).

52. Animals may also exploit visual illusions to make themselves either stand out from or blend into the background. Warning colors are intentionally conspicuous, in the same way Rembrandt used contrast to deceive our eyes. Conversely, countershading coloration—in which an animal's body has a dark back and a light underbelly—can make them hard to spot from both bottom up and top down, in the water and on land.

53. Gasparini, Serena, and Pilastro (2013).

54. Rather than using the colorless and tasteless word "foil" to refer to the uglier male guppy companion, Americans have accidentally credited the real inventor by using the word "stootfish," a slang word that you may find nowhere else but the Urban Dictionary.

55. Kelley and Endler (2012).

56. Macknik, Martinez-Conde, and Blakeslee (2011).

57. Abstract art here does not refer to the specific art school emerging in the beginning of the twentieth century with Kandinsky, Mondrian, and Malevich among the most famous. It refers generically to visual art with objects presented radically differently from our visual perception.

58. Singer et al. (2016).

59. Juslin and Västfjäll. (2008).

60. Brattico, Brattico, and Vuust (2017).

61. Brattico et al. (2016).

62. Endler (1992).

63. For a popular account of how the Fisherian runaway process can lead to exaggerated characteristics, see Prum (2017).

64. Kokko et al. (2002).

65. Some argue that for the idea of good genes to work, genetic correlations are necessary, but this condition is not stipulated in Zahavi's original hypothesis.

66. Recent studies show that a positive genetic link between a trait in males and the preference for the trait in females may not always be necessary for a runaway process. See Taylor and Ryan (2013).

67. Schmidt et al. (2017).

68. The legendary investment fund manager Cathie Wood predicted in 2021 that Bitcoin would reach $500,000.

69. Alain Sherter, "One Word Explains What Caused the Financial Crisis: Fraud," CBS News (website), May 5, 2010, https://www.cbsnews.com/news/one-word-explains-what-caused-the -financial-crisis-fraud/.

70. The then anonymous buyer was Saudi Culture Minister Prince Badr bin Abdullah.

71. Travis M. Andrews and Fred Barbash, "Long-lost da Vinci Painting Fetches $450.3 Million: An Auction Record for Art," *Washington Post*, November 16, 2017, https://www .washingtonpost.com/news/morning-mix/wp/2017/11/15/unimaginable-discovery-long-lost -da-vinci-painting-to-fetch-at-least-100-million-at-auction/.

72. Dutton (2009).

Chapter 6. Cheating in Humans

1. The movie was *inspired* by the autobiography (New York: Crown, 2000) but did not follow the details presented in the book. All information and quotes about Abagnale in this chapter are from the autobiography. Because I cite the book often, I have intentionally left out the page references.

2. DePaulo and Kashy (1998)

3. In a recent article, Abagnale claims that identity theft is 4,000 times easier to do now in our digital age. He says that he just needs two pieces of information—date and place of birth— to produce a fake identity. Unfortunately, these two pieces of information are provided voluntarily by many people in their digital social networks, such as Facebook.

4. Adams (1999).

5. Toma and Hancock (2010).

6. Treas and Giesen (2000). Whisman, Gordon, and Chatav (2007).

7. Wiederman (1997). Atkins, Baucom, and Jacobson (2001).

8. Tyler McCarthy, "Ashley Madison Hack Update," *International Business Times*, August 25, 2015, https://www.ibtimes.com/ashley-madison-hack-update-all-high-profile-celebrity-names -attached-private-2066211.

9. Petersen and Hyde (2010).

10. Rather than advertise their willingness for extra-marital affairs or brag about their sexual conquering as some men do, women tend to hide such information. Survey after survey show that women have markedly fewer sexual partners than men do. This even fooled researchers at times. But the numbers simply don't add up.

11. Arslan et al. (2018).

12. Jones, Hahn, and DeBruine (2019).

13. Bellis and Baker (1990).

14. Janus and Janus (1993).

15. Walum and Westberg (2011).

16. A caution for interpreting such genetic data. First, when a gene is responsible for 40% variation of a certain behavior, the result refers to a *statistical pattern* of a *population*, not any specific individual. Second, when the environmental or cultural conditions change, the value of genetic influence will also change. Since most genes work differently under different situations, genes seldom determine traits, especially behavioral traits.

17. Garcia et al. (2010)

18. Buss and Abrams (2017).

19. Larmuseau, Matthijs, and Wenseleers (2016).

20. Anderson (2006).

21. Scelza (2011).

22. Schmitt and Buss (2001).

23. Jankowiak, Nell, and Buckmaster (2002).

24. Daly and Wilson (1988). Betzig (1989). Goetz and Shackelford (2009).

25. Buss (2002).

26. At least in literature, if not so much in practice.

27. My paternal grandma was among the last generation of Chinese women with lotus feet. To avoid falling, she often had to hold onto walls while walking. Apparently, in the West small feet also hint at chastity, indicated, for instance, in the story of Cinderella.

28. Tess Sohngen, "11 Ridiculous, Sexist Laws that Still Exist in the US," Global Citizen (website), September 11, 2017, https://www.globalcitizen.org/en/content/sexist-laws-in-the-us -in-2017/.

29. Pazhoohi and Hosseinchari (2014). Pazhoohi (2016).

30. Onyishi et al. (2016).

31. Two examples of flattery: "You are *really* good!," when you are actually terrible; "We've made a major stride under your great leadership!," while the organization is falling apart.

32. We are socially trained to be agreeable—people generally prefer agreeable people— and being too critical or suspicious may alienate you from coworkers and friends. So cashiers and tellers normally won't offend customers by asking uncomfortable questions. Separately, but for the same reason, group brainstorming sessions are bad for generating diverse ideas. People, instead, look at one another and nod their heads to avoid differences in opinion.

33. Mortgage-backed securities were invented by Lewis Ranieri in the 1970s. They weren't broadly used as an investment vehicle until three decades later.

34. Faiss et al. (2020).

35. Here is another unwritten social rule: policing is normally unwelcome; doing it too often can guarantee you will become aliened from your colleagues and friends.

36. Bergdahl pleaded guilty before a military judge in 2017 and was punished with a reduction in rank, dishonorable discharge, and a $10,000 fine. All things considered, the reprimand Bergdahl received was quite lenient.

37. "Wells Fargo Account Fraud Scandal," Wikipedia, accessed May 18, 2022, https://en .wikipedia.org/wiki/Wells_Fargo_account_fraud_scandal.

38. Bagus and de Soto (2011).

39. Lynn (2010).

40. Ginsberg (2011).

41. Ginsberg (2011).

42. Toye (2006).

43. Weber (1968/1921). Hodson et al. (2013).

44. Jørgensen (2012).

45. Merton (1957).

46. If hiring is easy, firing is hard, especially in the public sector. In US federal agencies, for instance, new hires are on a probational basis for the first year. Once tenured, however, they are difficult to fire. Perhaps more galling is that even if they're incompetent at the job, their ranks and salaries will keep rising, quite comparable to parasites on juicy hosts.

47. Niskanen (1994). Carnis (2009).

48. Parkinson (1957).

49. Peter and Hull (1969).

50. Yolles (2016).

51. Kawai, Lang, and Li (2018).

52. "United States Intelligence Community," Wikipedia, accessed May 18, 2022, https://en .wikipedia.org/wiki/United_States_Intelligence_Community.

53. Jørgensen (2012).

54. Choi, Wiechman, and Pritchard (2013).

55. The conflict of interest and information asymmetry between a supervisor and a supervisee is generally known as a principal-agent problem.

56. Carpenter and Krause (2014).

57. Behn (1997).

58. Meier, O'Toole, and Bohte (2006).

59. Statistics show that up to 1% of government personnel consists of officials from presidential appointments, Schedule C jobs, and executive assignments; see Toye (2006).

60. Austrian economist Ludwig von Mises also held a similar view in his influential work *Bureaucracy*, published in 1944.

61. See the synthesis by Emily Gee and Topher Spiro, "Excess Administrative Costs Burden the US Health Care System," Center for American Progress (website), April 8, 2019, https:// www.americanprogress.org/issues/healthcare/reports/2019/04/08/468302/excess -administrative-costs-burden-u-s-health-care-system/.

62. Lockwood, Nathanson, and Weyl (2017).

63. Hodson et al. (2013).

64. Jacoby (2004).

65. Hodson et al. (2013).

66. The clip is available on YouTube: "Chinese Professor," Citizens Against Government Waste, October 20, 2010, YouTube video, 1:02, https://www.youtube.com/watch?v=OTSQ ozWP-rM.

67. Light (2017).

Chapter 7. Liars Who Lie to Themselves

1. The Greek philosopher Thales is said to be the first to use the term "Know Thyself."

2. Hoorens and Harris (1998).

3. Svenson (1981).

4. Alicke and Govorun (2005).

5. Zuckerman, Ezra, and Jost (2001).

6. Cross (1977). Zuckerman, Ezra, Jost (2001).

7. Neale and Bazerman (1985). Odean (1998).

8. Personal identity is a philosophical and psychological issue. For the Enlightenment philosopher John Locke, it refers to the consciousness of being the same person over time, connected through memory and self-awareness. Thus, if you can't remember your past or do not think the past you is the same you now, you are indeed a new person.

9. Trivers (2011).

10. Epley and Whitchurch (2008).

11. Epley and Whitchurch (2008).

12. Anecdotes of self-deception have been reported in many nonhuman animals, but decisive evidence is still largely lacking, due mainly to the difficulty of demonstrating that animals have knowledge about themselves. Even so, a recent study has provided positive evidence for self-deception in male crayfish; see Anguilletta, Kubitz, and Wilson (2019).

13. Kruger and Dunning (1999).

14. Sometimes, it is also known as the Lake Wobegon effect, for the obvious reason.

15. Kruger and Dunning's article won the Ig Nobel Prize, a prankish award given to published research with outrageously funny or surprising findings. The prize is partly for spicing up the overly serious academic life and partly for discoveries whose implications are deeper than we can fully grasp for the time being. Clearly, Kruger and Dunning's award-winning research belongs to the latter category.

16. Dunning (2011).

17. Demanding mental work can drain our willpower and alter our decisions, a phenomenon known as ego depletion among psychologists. For instance, when working on a hard task, people are more likely to eat cookies against their promise of only eating healthy food. In a study in Israel, judges handed out favorable parole decisions (65% of the time) immediately after a meal, which then gradually dropped to nearly 0% before the next meal; see Danziger et al. (2011).

18. There are numerous video clips available on the Internet, including: John Shirek, Hope Ford, Johnathan Raymond, "'She was my baby': Father Spoke in Past Tense of Missing 2-week-old before She Was Found Dead," 11 Alive (website), May 10, 2019, https://www.11alive

.com/article/news/father-spoke-of-missing-2-week-old-in-past-tense-while-alone-with
-mother-in-interrogation-room/85-70472a8a-c9b5-4a8a-910c-5b0fd62503dc.

19. Alexis Stevens, "Newton County Parents Guilty in 2-week-old's Death," *Atlanta Journal Constitution*, May 14, 2019, https://www.ajc.com/news/crime--law/breaking-newton-county
-parents-guilty-week-old-death/uJmGfBI0BhSCdWUpKzmigO/.

20. Trivers (2011). Von Hippel and Trivers (2011).

21. Trivers (2011).

22. Kwan et al. (2007).

23. Suls, Lemos, and Stewart (2002).

24. Plassmann et al. (2008).

25. "Global warming" and "climate change" were coined in 1975 and 1979, respectively. Today, "climate change" is more broadly used due partly to the political effort by the George W. Bush administration to tone down the implied crisis of global warming and partly to the inclusive nature of the term, which can refer to both a globally warming climate and its subsequent regional weather changes, which can be erratic.

26. Ditto and Lopez (2003).

27. Mather and Carstensen (2005)

28. Tavris and Aronson (2015).

29. D'Argembeau and van der Linden (2008).

30. Loftus and Pickrell (1995).

31. Howe and Knott (2015).

32. Schreiber et al. (2006).

33. Trivers (2011).

34. This is a partial list from Walton (2019).

35. Politicians are often stereotyped as or blamed for being dishonest, but can the voting crowd be free of guilt for that?

36. Although bumping into a poisonous snake is mostly a low-probability event, it's virtually a certainty in your life if you live in a snake-infested forest like tribal people in Africa.

37. Galperin and Haselton (2012).

38. In psychology and psychiatry, perceiving nonexisting patterns is known as pareidolia, whereas connecting two random events is termed apophenia.

39. I've never figured out what it was and am glad that the stuff was not highly toxic.

40. Bingel et al. (2011).

41. Benedetti, Carlino, and Pollo (2011).

42. Benedetti and Piedimonte (2019).

43. Fournier et al. (2010).

44. Charlesworth et al. (2017).

45. Price, Finniss, and Benedetti (2008). Benedetti (2009).

46. De Craen et al. (1996).

47. Kaptchuk and Miller (2015).

48. Benedetti, Carlino, and Pollo (2011).

49. Wager et al. (2004).

50. Scott et al. (2007).

51. Benedetti (2010).

52. Science advances by the powerful method of falsification, through which truths accumulate while falsehoods are culled. Alternative medicine does not follow this. As a result, its body of knowledge does not progress as science does. Chinese medicine, for instance, has not made a major theoretical breakthrough since *The Yellow Emperor's Classic of Internal Medicine* (or *Huangdi Neijing*) was compiled more than two thousand years ago. Even today, standard scientific procedures such as randomization with double-blind trials are still not rigorously used in many clinical trials in traditional Chinese medicine.

53. McGeeney (2015).

54. Linde et al. (2005). Linde et al. (2007).

55. Finniss et al. (2010).

56. McGeeney (2015).

57. Kaptchuk and Miller (2015).

58. Kaptchuk and Miller (2015).

59. Finniss et al. (2010).

60. Depressed people tend to have a more realistic view about the world, including a sense of losing control of their lives; see Alloy and Clements (1992). But even so, they still lean toward being optimistic, unlike true pessimists.

61. Dufner et al. (2012).

62. Carver, Scheier, and Segerstrom (2010).

63. Bishop, Tuchfarber, and Oldendick (1986). Graeff (2003). Paulhus et al. (2003).

64. Atir, Rosenzweig, and Dunning (2015).

65. Darwin (1871).

66. Paulhus et al. (2003).

67. Rozenblit and Keil (2002).

68. Lusardi and Mitchell (2009).

69. Vnuk, Owen, and Plummer (2006).

70. Trivers (2011).

71. Trivers (2011).

72. Chatterjee and Hambrick (2011).

73. Eisenegger et al. (2017).

74. Kamiya, Kim, and Suh (2016).

75. Dawson, Savitsky, and Dunning (2006).

76. "Cryptoqueen: How this Woman Scammed the World, then Vanished," *BBC News*, November 24, 2019, https://www.bbc.com/news/stories-50435014.

77. His name is Igor Alberts, who claimed to have accumulated €100 million from his business. By the way, the pyramid scam and its variants are not illegal in many countries. Herbalife, a publicly traded company listed on the New York Stock Exchange, appears to be a business based on this model.

78. Vosoughi, Roy, and Aral (2018).

79. Grinberg et al. (2019).

80. "The Disinformation Dozen," Center for Countering Digital Hate (website), March 21, 2021, https://www.counterhate.com/disinformationdozen.

81. Nyhan and Reifler (2010).

82. Bail et al. (2018).

83. Westen et al. (2006).

84. Greenberg, Solomon, and Pyszczynski (1997).

85. Albarracín and Mitchell (2004). Kumashiro and Sedikides (2005).

86. Answers: 1. Ernest Hemingway; 2. Gandhi; 3. Ralph Waldo Emerson; 4. Albert Einstein; 5. Lao Tzu; 6. Alexander Pope; 7. Benjamin Franklin; 8. Confucius. Note that overconfidence is more of a problem for men than for women for reasons we will see soon. Unsurprisingly, most memorable quotes for staying humble are from men.

87. Armitage et al. (2008).

88. One way to overcome the confirmation bias is through a method called motivational interviewing in psychological consulting. Through open-minded questioning, people are given an opportunity to rediscover themselves including reexamining their preconceived views.

89. Kaufmann (2008).

90. Ehrlinger and Dunning (2003).

91. Hoobler et al. (2016).

92. In Tibetan macaques, we found that when more females are involved, the speed and maybe accuracy of collective decisions will go up. Apparently, better social connections among females facilitate the decision-making process; see Fratellone et al. (2019). It is difficult, though, to know whether female macaques are less likely than males to be overconfident.

93. Hoobler et al. (2016).

Chapter 8. Living with Lies and Deceptions

1. Tenbmnsel (1998).

2. Google eventually bought YouTube for $1.65 billion. YouTube generated $15.15 billion in advertisement revenue for Google in fiscal year 2019 alone.

3. Thaler and Sunstein. 2008.

4. Gneezv (2005).

5. Friedman (1970).

6. Murphey, Laczniak, and Wood (2007). Brenkert (1999).

7. Borgerson and Schroeder (2008).

8. This raises a whole new question for debate. Should we be honest when the justice system is unjust? In Victor Hugo's epic novel *Les Misérables*, Javert, the police inspector, jumps into the Seine and ends his life for losing his faith in a justice system he has fanatically defended.

9. Social conventions may overrule laws. For instance, people drive 5–9 miles above the speed limit in many places. If you choose to comply with the law, it's likely that you may cause traffic jams and be cursed by other drivers. Often, it's dangerous if you do not go with the traffic flow.

10. Philosophers may raise the question: Who is the real me—the public me in a suit or the private me in a T-shirt? The answer is that I am both, one better in appearance than the other. This double life illustrates how important and necessary it is to present a better public image of oneself by covering up one's weaknesses.

11. People may habitually make themselves look good. Selectively presenting a better image in public, even without a clear intention to do so, is by definition not fully honest. This is comparable with beautifying a digital image by airbrushing to appeal to viewers.

12. Far worse than cherry-picking, Gladwell has been called out by multiple news media for plagiarism in his writings since 2010.

13. Malcolm Gladwell, "Christopher Chabris Should Calm Down," *Slate*, October 10, 2013, https://slate.com/technology/2013/10/malcolm-gladwells-david-and-goliath-he-explains -why-christopher-chabris-criticisms-of-his-book-were-unreasonable.html.

14. Nyberg (1993).

15. Smith (2007). The quote within this quote is from: R. D. Alexander, "The Search for a General Theory of Behavior," *Behavioral Science* 10 (1975): 96.

16. Taglor (2007).

17. Reddy (2007).

18. Xu et al. (2010).

19. Talwar and Crossman (2011).

20. Lewis (1993).

21. Carlson, Moses, and Hix (1998). Talwar and Lee (2008).

22. Sodian and Frith (1992).

23. Reddy (2007).

24. Dor (2017).

25. Talwar and Crossman (2011).

26. Kant (1797).

27. Weinrib (2008).

28. Kant (1797)

29. Melville (2014).

30. Most of us believe in consequentialism to a certain degree, so we think you can lie to protect the friend. Kant, however, believes that you should care about doing right, regardless of the consequences. He thinks that we have a moral duty not to lie and that the wrongness of lying is not because of its consequences. He has two basic arguments for why lying is wrong. First, we couldn't universalize it. If everyone lied, then people wouldn't believe me when I lied. I wouldn't accomplish the aim of my lie. The lie only works if people have a general sense of trust. In lying, then, I am making an exception of myself. I am saying that everyone else should generally tell the truth, but I'm going to lie. That's irrational. Second, I am using the people I'm lying to merely as a means. I am manipulating or coercing them and thus not treating them as individuals who can make their own decisions.

31. Zupancic (2000).

32. Constant (1988).

33. Kant (1797).

34. Carson (2012).

35. Listed in "Lie," Wikipedia, https://en.wikipedia.org/wiki/Lie; and based on Augustine's two books: "On Lying" (*De Mendacio*) and "Against Lying" (*Contra Mendacio*).

36. This doesn't mean that moral principles matter little, however. On the contrary, most of us make moral decisions in practice by a combination of principles and consequences.

37. Most political and policy changes in a democracy will hurt some people, usually minorities, leading to oxymoronic situations known as the dictatorship of the majority. This dilemma can't be resolved by pure utilitarianist approaches. As such, the stipulation of some (deontological) rules are vital in addressing minority concerns.

38. She was acquitted for the charge of insurance fraud and allowed to keep her job for her track record of Good Samaritan acts in the community.

39. In cases like this, Chinese physicians must disclose the truthful information to the patient's close relatives, who help make decisions for the ensuing medical treatment. Some decades ago, American physicians practiced the same to a certain degree, but do not do so today due to legal complications.

40. This issue will unleash the debate between cultural relativists, who believe that all cultural aspects should be respected, and cultural absolutists, who think that certain principles and values are objectively right regardless of social conditions.

41. With about 60,000 residents at the time, the port city of Königsberg was by no means a closed city. The commercial life and social dynamism of the city, however, might have little to do with the philosopher who lived an extremely private life as a single man.

42. Harris (2013).

43. Cohn et al. (2019).

44. Kar and Freitas (2009).

45. Gachter and Schulz (2016).

46. Norris, Brookes, and Dowell (2019).

47. Lixing Sun, "Would Twitter Ruin Bee Democracy?," *Nautilus*, December 14, 2017, https://nautil.us/issue/55/trust/would-twitter-ruin-bee-democracy.

48. Vilmer et al. (2018).

49. The last time Congress had to be evacuated was during the War of 1812.

50. Frankel (2012).

51. Van der Linden et al. (2017).

52. As of June 15, 2020, CrowdStrike's market value was $21.72 billion, a gain of more than 90% in a year. In the following year, its valuation increased by 137%; as of June 15, 2021, its market value was $51.48 billion.

53. Steve Morgan, "Global Cybersecurity Spending Predicted to Exceed $1 Trillion from 2017–2021," *Cybercrime Magazine*, June 10, 2019, https://cybersecurityventures.com/cybersecurity-market-report/.

54. Clarke and Knake (2019).

55. Hegel (1821). There are several translated versions of the same statement. Its real meaning is still a subject of debate.

BIBLIOGRAPHY

Adams, M. (1999). The dead grandmother/exam syndrome. *Annals of Improbable Research* 5: 3–6.

Akino, T., Knapp, J. J., Thomas, J. A., and Elmes, G. W. (1999). Chemical mimicry and host specificity in the butterfly *Maculinea rebeli*, a social parasite of Myrmica ant colonies. *Proceedings of the Royal Society B* 266: 1419–1426.

Albarracín, D., and Mitchell, A. L. (2004). The role of defensive confidence in preference for proattitudinal information: How believing that one is strong can sometimes be a defensive weakness. *Personality and Social Psychology Bulletin* 30: 1565–1584.

Alicke, M. D., and Govorun, O. (2005). The better-than-average effect. In Alicke, M. D., Dunning, D. A., Krueger, J. I., eds., *The Self in Social Judgment (Studies in Self and Identity)*, 85–106. New York: Psychology Press.

Allies, A. B., Bourke, A.F.G., and Franks, N. R. (1986). Propaganda substances in the cuckoo ant *Leptothorax kutteri* and the slave-maker *Harpagoxenus sublaevis*. *Journal of Chemical Ecology* 12: 1285–1293.

Alloy, L. B., and Clements, C. M. (1992). Illusion of control: Invulnerability to negative affect and depressive symptoms after laboratory and natural stressors. *Journal of Abnormal Psychology* 101: 234–245.

Als, T. D., Vila, R., Kandul, N. P., Nash, D. R., Yen, S.-H., Hsu, Y.-F., Mignault, A. A., Boomsma, J. J., and Pierce, N. E. (2004). The evolution of alternative parasitic life histories in large blue butterflies. *Nature* 432: 386–390.

Anderson, J. R., Kuroshima, H., Kuwahata, H., Fujita, K., and Vick, S. (2001). Training squirrel monkeys (*Saimiri sciureus*) to deceive: Acquisition and analysis of behaviour toward cooperative and competitive trainers. *Journal of Comparative Psychology* 115: 282–293.

Anderson, K. G. (2006). How well does paternity confidence match actual paternity? Evidence from worldwide nonpaternity rates. *Current Anthropology* 47: 513–520.

Andrews, T. M., Lukaszewski, A. W., Simmons, Z. L., and Bleske-Rechek, A. (2017). Cue-based estimates of reproductive value explain women's body attractiveness. *Evolution and Human Behavior* 38: 461–467.

Anguilletta Jr., M. J., Kubitz, G., and Wilson, R. S. (2019). Self-deception in nonhuman animals: Weak crayfish escalated aggression as if they were strong. *Behavioral Ecology* 30: 1469–1476.

Armitage, C. J., Harris, P. R., Hepton, G., and Napper, L. (2008). Self-affirmation increases acceptance of health-risk information among UK adult smokers with low socioeconomic status. *Psychology of Addictive Behaviors* 22: 88–95.

Arnqvist, G., and Kirkpatrick, M. (2005). The evolution of infidelity in socially monogamous passerines: The strength of direct and indirect selection on extrapair copulation behavior in females. *American Naturalist* 165: S26–S37.

Arslan, R. C., Schilling, K. M., Gerlach, T. M., and Penke, L. (2021). Using 26,000 diary entries to show ovulatory changes in sexual desire and behavior. *Journal of Personality and Social Psychology* 121: 410–431.

Ashton, B. J., Thornton, A., and Ridley, A. R. (2018). An intraspecific appraisal of the social intelligence hypothesis. *Philosophical Transactions of the Royal Society B* 373: 20170288.

Atir, S., Rosenzweig, E., and Dunning, D. (2015). When knowledge knows no bounds: Self-perceived expertise predicts claims of impossible knowledge. *Psychological Science* 26: 1295–1303.

Atkins, D. C., Baucom, D. H., and Jacobson, N. S. (2001). Understanding infidelity: Correlates in a national random sample. *Journal of Family Psychology* 15: 735–749.

Baglione, V., Canestrari, D., Chiarati, E., Vera, R., and Marcos, J. M. (2010). Lazy group members are substitute helpers in carrion crows. *Proceedings of the Royal Society B* 277: 3275–3282.

Bagus, P., and de Soto, J. H. (2011). *The Tragedy of the Euro*. Auburn, AL: Ludwig von Mises Institute.

Bail, C. A., Argyle, L. P., Brown, T. W., Bumpus, J. P., Chen, H., Hunzaker, M.B.F., Lee, J., Mann, M., Merhout, F., and Volfovsky, A. (2018). Exposure to opposing views on social media can increase political polarization. *Proceedings of the National Academy of Sciences* 115: 9216–9221.

Barber, J. R., and Conner, W. E. (2007). Acoustic mimicry in a predator-prey interaction. *Proceedings of the National Academy of Sciences* 104: 9331–9334.

Barbero, F., Thomas, J. A., Bonelli, S., Balletto, E., and Schonrogge, K. (2009). Queen ants make distinctive sounds that are mimicked by a butterfly social parasite. *Science* 323: 782–785.

Barclay, P., and Willer, R. (2006). Partner choice creates competitive altruism in humans. *Proceedings of the Royal Society B* 274: 749–752.

Barrett, L., and Henzi, P. (2005). Social nature of cognition. *Proceedings of the Royal Society B* 272: 1865–1875.

Barsh, G. (2016). Evolution: Sex, diet and red ketocarotenoids. *Current Biology* R1145–R1147.

Basolo, A. (1990). Female preference predates the evolution of the sword in swordtail fish. *Science* 250: 808–810.

Bass, A. H. (1996). Shaping brain sexuality. *American Scientist* 84: 352–364.

Batzer, M. A., and Deininger, P. L. (2002). Alu repeats and human genome diversity. *Nature Reviews Genetics* 3: 370–329.

Bauer, U., Federle, W., Seidel, H., Grafe, U., and Ioannou, C. (2015). How to catch more prey with less effective traps: Explaining the evolution of temporarily inactive traps in carnivorous pitcher plants. *Proceeding of the Royal Society B* 282: 2675.

Bee, M. A., Perrill, S. A., and Owen, P. C. (2000). Male green frogs lower the pitch of acoustic signals in defense of territories: A possible dishonest signal of size? *Behavioral Ecology* 11: 169–177.

Beekman, M., and Oldroyd, B. P. (2008). When workers disunite: Intraspecific parasitism by eusocial bees. *Annual Review of Entomology* 53: 19–37.

Behn, R. (1997). Linking measurement to motivation. *Advances in Educational Administration* 5: 15–50.

Bell, E., and Buchner, A. (2012). How adaptive is memory for cheaters? *Current Directions in Psychological Science* 21: 403–408.

Bellis, M. A., and Baker, R. R. (1990). Do females promote sperm competition? Data for humans. *Animal Behaviour* 40: 997–999.

Benedetti, F. (2009). *Placebo Effects: Understanding the Mechanisms in Health and Disease*. New York: Oxford University Press.

———. (2010). No prefrontal control, no placebo response. *Pain* 148: 357–358.

Benedetti, F., Carlino, E., and Pollo, A. (2011). How placebos change the patient's brain. *Neuropsychopharmacology Reviews* 36: 339–354.

Benedetti, F., and Piedimonte, A. (2019). The neurobiological underpinnings of placebo and nocebo effects. *Seminars in Arthritis and Rheumatism* 49: S18–S21.

Bereczkei, T., Papp, P., Kincses, P., Bodrogi, B., Perlaki, G., Orsi, G., and Deak, A. (2015). The neural basis of the Machiavellians' decision making in fair and unfair situations. *Brain and Cognition* 98: 53–64.

Betzig, L. (1989). Causes of conjugal dissolution: A cross-cultural study. *Current Anthropology* 30: 654–676.

Bingel, U., Wanigasekera, V., Wiech, K., Mhuircheartaigh, R. N., Lee, M. C., Ploner, M., and Tracey, I. (2011). The effect of treatment expectation on drug efficacy: Imaging the analgesic benefit of the opioid remifentanil. *Science Translational Medicine* 3: 70ra14.

Bird, R. B., and Smith, E. A. (2005). Signaling theory, strategic interaction, and symbolic capital. *Current Anthropology* 46: 221–248.

Bishop, G. F., Tuchfarber, A. J., and Oldendick, R. W. (1986). Opinions on fictitious issues: The pressure to answer survey questions. *Public Opinion Quarterly* 50: 240–250.

Borgerson, J. L., and Schroeder, J. E. (2008). Building an ethics of visual representation: Contesting epistemic closure in marketing communication. In Morland, M. P., and Werhane, P., eds., *Cutting Edge Issues in Business Ethics*, 87–108. Boston: Springer.

Bourke, A.F.G. (2011). *Principles of Social Evolution*. Oxford: Oxford University Press.

Brattico, E., Bogert, B., Alluri, V., Tervaniemi, M., Eerola, T., and Jacobsen, T. (2016). It's sad but I like it: The neural dissociation between musical emotions and liking in experts and laypersons. *Frontiers in Human Neuroscience* 9: 676.

Brattico, P., Brattico, E., and Vuust, P. (2017). Global sensory qualities and aesthetic experience in music. *Frontiers in Neuroscience* 11: 159.

Brenkert, G. K. (1999). Marketing ethics. In Frederick, R. E., ed., *A Companion to Business Ethics*, 178–197. Malden, MA: Blackwell.

Bruce, H. M. (1959). An exteroceptive block to pregnancy in the mouse. *Nature* 184: 4680.

Bruce, J. B., Cooper, G. A., Chabas, H., West, S. A., and Griffin, A. S. (2017). Cheating and resistance to cheating in natural populations of the bacterium *Pseudomonas fluorescens*. *Evolution* 71: 2484–2495.

Bshary, R. (2002). Biting cleaner fish use altruism to deceive image-scoring client reef fish. *Proceedings of Royal Society of London B* 269: 2087–2093.

Bugnyar, T., and Kotrschal, K. (2002). Observational learning and the raiding of food caches in ravens, *Corvus corax*: Is it 'tactical' deception? *Animal Behaviour* 64: 185–195.

———. (2004). Leading a conspecific away from food in ravens (*Corvus corax*)? *Animal Cognition* 7: 69–76.

Burley, N. (1988). Wild zebra finches have band-colour preferences. *Animal Behaviour* 36: 1235–1237.

Burley, N. T., and R. Symanski. (1998). "A taste for the beautiful": Latent aesthetic mate preferences for white crests in two species of Australian grassfinches. *American Naturalist* 152: 792–802.

Burt, A., and Trivers, R. L. (2006). *Genes in Conflict: The Biology of Selfish Genetic Elements*. Cambridge, MA: Belknap Press of Harvard University Press.

Buss, D. (2002). Human mate guarding. *Neuroendocrinology Letters* 23(Suppl.4): 23–29.

Buss, D. M., and Abrams, M. (2017). Jealousy, infidelity, and the difficulty of diagnosing pathology: A CBT approach to coping with sexual betrayal and the green-eyed monster. *Journal of Rational-Emotive and Cognitive-Behavior Therapy* 35: 150–172.

Butaitė, E., Baumgartner, M., Wyder, S., and Kümmerli, R. (2016). Siderophore cheating and cheating resistance shape competition for iron in soil and freshwater *Pseudomonas* communities. *Nature Communications* 8: 414.

Byrne, R. W. (2018). Machiavellian intelligence retrospective. *Journal of Comparative Psychology* 132: 432–436.

Byrne, R. W., and Corp, N. (2004). Neocortex size predicts deception rate in primates. *Proceedings of the Royal Society B* 271: 1693–1699.

Byrne, R. W., and Whiten, A. (1985). Tactical deception of familiar individuals in baboons (*Papio ursinus*). *Animal Behaviour* 33: 669–673.

———. (1990). Tactical deception in primates: The 1990 database. *Primate Report* 27: 1–101.

———. (1992). Cognitive evolution in primates: Evidence from tactical deception. *Man* (New Series) 27: 609–627.

Carlson, S. M., Moses, L. J., and Hix, H. R. (1998). The role of inhibitory control in young children's difficulties with deception and false belief. *Child Development* 69: 672–691.

Carnis, L.A.H. (2009). The economic theory of bureaucracy: Insights from the Niskanian model and Misesian approach. *Quarterly Journal of Austrian Economics* 12: 57–78.

Caro, T. (2016). *Zebra Stripes*. Chicago: University of Chicago Press.

Caro, T., Izzo, A., Reiner, R. C., Walker, H., and Stankowich, T. (2014). The function of zebra stripes. *Nature Communications* 5: 3535.

Carpenter, D., and Krause, G. A. (2014). Transactional authority and bureaucratic politics. *Journal of Public Administration Research and Theory* 25: 5–25.

Carson, T. L. (2012). *Lying and Deception: Theory and Practice*. Oxford: Oxford University Press.

Carter, G. G., and Wilkinson, G. S. (2013). Food sharing in vampire bats: Reciprocal help predicts donations more than relatedness or harassment. *Proceedings of the Royal Society B* 280: 20122573.

Carver, C. S., Scheier, M. F., and Segerstrom, S. C. (2010). Optimism. *Clinical Psychology Review* 30: 879–889.

Charlesworth, J. E., Petkovic, G., Kelley, J. M., Hunter, M., Onakpoya, I., Roberts, N., Miller, F. G., and Howick, J. (2017). Effects of placebos without deception compared with no treatment: A systematic review and meta-analysis. *Journal of Evidence-Based Medicine* 10: 97–107.

Chatterjee, A., and Hambrick, D. C. (2011). Executive personality, capability cues, and risk taking: How narcissistic CEOs react to their successes and stumbles. *Administrative Science Quarterly* 56: 202–237.

Chen, J., Zou, Y., Sun, Y.-H., and ten Cate, C. (2019). Problem-solving males become more attractive to female budgerigars. *Science* 363: 166–167.

Cheney, K. L. (2012). Cleaner wrasse mimics inflict higher costs on their models when they are more aggressive towards signal receivers. *Biology Letters* 8: 10–12.

Choi, S. J., Wiechman, A. C., and Pritchard, A. C. (2013). Scandal enforcement at the SEC: The arc of the option backdating investigations. *American Law and Economics Review* 15: 542–577.

Christy, J. H. (1995). Mimicry, mate choice, and the sensory trap hypothesis. *American Naturalist* 146: 171–81.

Clarke, R. A., and Knake, R. K. (2019). *The Fifth Domain*. New York: Penguin.

Cloud, J. M., and Taylor, M. H. (2019). The effect of mate value discrepancy on hypothetical engagement ring purchases. *Evolutionary Psychological Science* 5: 22–28.

Cohn, A., Maréchal, M. A., Tannenbaum, D., and Zünd, C. L. (2019). Civic honesty around the globe. *Science* 362: 70–73.

Colombelli-Négrel, D., Hauber, M. E., Robertson, J., Sulloway, F. J., Hoi, H., Griggio, M., and Kleindorfer, S. (2012). Embryonic learning of vocal passwords in superb fairy-wrens reveals intruder cuckoo nestlings. *Current Biology* 20: 2155–2160.

Constant, B. (1988). Des Réactions Politiques. In Constant, B., ed. *De La Force du Gouvernement Actuel de la France*. Paris: Flammarion. Cited in Rousseliere, G. (2018). On political responsibility in post-revolutionary times: Kant and Constant's debate on lying. *European Journal of Political Theory* 17: 214–232.

Cosmides, L., and Tooby, J. (1992). Adaptations for social exchange. In Barkow, J. H., Cosmides, L., and Tooby, J., eds, *The Adapted Mind: Evolutionary Psychology and the Generation of Culture*, 163–228. New York: Oxford university Press. (For a more generic introduction, see Christopher Badcock, "Making Sense of Wason," *Psychology Today* (blog), May 5, 2012, www .psychologytoday.com/us/blog/the-imprinted-brain/201205/making-sense-wason.)

Coussi-Korbel, S. (1994). Learning to outwit a competitor in mangabeys (*Cercocebus torquatus torquatus*). *Journal of Comparative Psychology* 108: 164–171.

Crawford, J. C., Liu, Z., Nelson, T. A., Nielsen, C. K., and Bloomquist, C. K. (2008). Microsatellite analysis of mating and kinship in beavers (*Castor canadensis*). *Journal of Mammalogy* 89: 575–581.

Crews, D., and Garstika, W. R. (1982). The ecological physiology of a garter snake. *Scientific American* 11: 159–168.

Cross, K. P. (1977). Not can but will college teachers be improved? *New Directions for Higher Education* 17: 1–15.

Cui, J., Tang, Y., and Narin, P. M. (2012). Real estate ads in Emei music frog vocalizations: Female preference for calls emanating from burrows. *Biology Letters* 8: 337–340.

D'Argembeau, A., and van der Linden, M. (2008). Remembering pride and shame: Self-enhancement and the phenomenology of autobiographical memory. *Memory* 16: 538–547.

Daly, M., and Wilson, M. (1988). Evolutionary social psychology and family homicide. *Science* 242: 519–524.

Frankel, T. (2012). *The Ponzi Scheme Puzzle: A History and Analysis of Con Artists and Victims.* New York: Oxford University Press.

Fratellone, G. P., Li, J. H., Sheeran, L. K., Wagner, R. S., Wang, X., and Sun, L. (2019). Social connectivity among female Tibetan macaques (*Macaca thibetana*) increases the speed of collective movements. *Primates* 60: 183–189.

Friedman, M. (1970). The Social Responsibility of Business is to Increase Its Profits. *New York Times Magazine,* September 13.

Gachter, S., and Schulz, J. F. (2016). Intrinsic honesty and the prevalence of rule violations across societies. *Nature* 531: 496–499.

Galperin, A., and Haselton, M. G. (2012). The evolution of cognitive bias. In Forgas, J., Fiedler, K., and Sedikedes, C., eds., *Social Thinking and Interpersonal Behavior,* 45–64. New York: Psychology Press.

Garcia, J. R., MacKillop, J., Aller, E. L., Merriwether, A. M., Wilson, D. S., and Lum, J. K. (2010). Associations between dopamine D4 receptor gene variation with both infidelity and sexual promiscuity. *PLoS ONE* 5: e14162.

Gasparini, C., Serena, G., and Pilastro, A. 2013. Do unattractive friends make you look better? Context-dependent male mating preferences in the guppy. *Proceedings of the Royal Society B* 280: 3072.

Gerhardt, H. C., and Huber, F. (2002). *Acoustic Communication in Insects and Anurans.* Chicago: University of Chicago Press.

Gerlach, N. M., McGlothlin, J. W., Parker, P. G., and Ketterson, E. D. (2012). Promiscuous mating produces offspring with higher lifetime fitness. *Proceedings of the Royal Society B* 279: 860–866.

Ghoul, M., Griffin, A. S., and West, S. A. (2014). Toward an evolutionary definition of cheating. *Evolution* 68: 318–331.

Gilbert, L. E. (1982). The co-evolution of a butterfly and a vine. *Scientific American* 247: 110–121.

Gilot, F., and Lake, C. (2019). *Life with Picasso.* New York: NYRB Classics, p. 266.

Ginsberg, B. (2011). *The Fall of the Faculty.* Oxford: Oxford University Press.

Gneezv, U. (2005). Deception: The role of consequences. *The American Economic Review* 95: 384–394.

Goetz, A. T., and Shackelford, T. K. (2009). Sexual coercion in intimate relationships: A comparative analysis of the effects of women's infidelity and men's dominance and control. *Archives of Sexual Behavior* 38: 226–234.

Gonçalves, B., Perra, N., and Vespignani, A. (2011). Modeling users' activity on Twitter networks: Validation of Dunbar's number. *PLoS ONE* 6: e22656.

Goodall, J. (1986). *The Chimpanzees of Gombe.* Cambridge, MA: Belknap Press.

Gopnik, A. (1993). How we know our minds: The illusion of first-person knowledge of intentionality. In Goldman, A. I., ed., *Readings in Philosophy and Cognitive Science,* 315–346. Cambridge, MA: MIT Press.

Graeff, T. R. (2003). Exploring consumers' answers to survey questions: Are uninformed responses truly uninformed? *Psychology and Marketing* 20: 643–667.

Graphodatsky, A. S., Trifonov, V. A., and Stanyon, R. (2011). The genome diversity and karyotype evolution of mammals. *Molecular Cytogenetics* 4: 22.

Greenberg, J., Solomon, S., and Pyszczynski, T. (1997). Terror management theory of self-esteem and cultural worldviews: Empirical assessments and conceptual refinements. *Advances in Experimental Social Psychology* 29: 61–139.

Greitemeyer, T., Kastenmüller, A., and Fischer, P. (2013). Romantic motives and risk-taking: An evolutionary approach. *Journal of Risk Research* 16: 19–38.

Griffin, A. S., West, S. A., and Buckling, A. (2004). Cooperation and competition in pathogenic bacteria. *Nature* 430: 1024–1027.

Griffith, S. C., Owens, I.P.F., and Thuman, K. A. (2002). Extra pair paternity in birds: A review of interspecific variation and adaptive function. *Molecular Ecology* 11: 2195–2212.

Grinberg, N., Joseph, K., Friedland, L., Swire-Thompson, B., and Lazer, D. (2019). Fake news on Twitter during the 2016 U.S. presidential election. *Science* 363: 374–378.

Hafernik, J., and Saul-Gershenz, L. S. (2000). Beetle larvae cooperate to mimic bees. *Nature* 405: 35.

Hagelin, J. C. (2002). The kinds of traits involved in male-male competition: A comparison of plumage, behavior, and body size in quail. *Behavioral Ecology* 13: 32–41.

Hamilton, W. D., and Zuk, M. (1982). Heritable true fitness and bright birds: A role for parasites? *Science* 218: 384–387.

Hancks, D. C., and Kazazian Jr., H. H. (2016). Roles for retrotransposon insertions in human disease. *Mobile DNA* 7: 9.

Hanlon, R. T., Forsythe, J. W., and Joneschild, D. E. (1999). Crypsis, conspicuousness, mimicry, and polyphenism as antipredator defenses of foraging octopuses on Indo-Pacific coral reefs, with a method of quantifying crypsis from video tapes. *Biological Journal of the Linnaean Society* 66: 1–22.

Hanlon, R. T., Naud, M. J., Shaw, P. W., and Havenhand, J. N. (2005). Transient sexual mimicry leads to fertilization. *Nature* 430: 212.

Hare, J. F., and Atkins, B. A. (2001). The squirrel that cried wolf: Reliability detection by juvenile Richardson's ground squirrels (*Spermophilus recharsonii*). *Behavioral Ecology and Sociobiology* 51: 108–112.

Harris, S. (2013). *Lying*. Cleveland, OH: Four Elephants Press.

Hauser, M. D. (1992). Costs of deception: cheaters are punished in rhesus monkeys (*Macaca mulatta*). *Proceedings of the National Academy of Sciences* 89: 12137–12139.

Hegel, G.W.F. (1821). The Preface to *Elements of the Philosophy of Right* (*Philosophie als Wissenschaft*). Berlin: De Gruyter. (There are several translated versions of the same statement. Its real meaning is still a subject of debate).

Heinsohn, R., and Packer, C. (1995). Complex cooperative strategies in group-territorial African lions. *Science* 269: 1260–1262.

Hirata, S., and Matsuzawa, T. (2001). Tactics to obtain a hidden food item in chimpanzee pairs (*Pan troglodytes*). *Animal Cognition* 4: 285–295.

Hodson, R., Roscigno, V. J., Martin, A., and Lopez, S. H. (2013). The ascension of Kafkaesque bureaucracy in private sector organization. *Human Relations* 66: 1249–1273.

Hoobler, J. M., Masterson, C. R., Nkomo, S. M., and Michel, E. J. (2018). The business case for women leaders: Meta-analysis, research critique, and path forward. *Journal of Management* 44: 2473–2499.

Hoorens, V., and Harris, P. (1998). Distortions in reports of health behaviours: The time span effect and illusory superiority. *Psychology and Health* 13: 451–466.

Hoover, J. P., and Robinson, K. (2007). Retaliatory mafia behavior by a parasitic cowbird favors host acceptance of parasitic eggs. *Proceedings of the National Academy of Sciences* 104: 4479–4483.

Howe, M. L., and Knott, L. M. (2015). The fallibility of memory in judicial processes: Lessons from the past and their modern consequences. *Memory* 23: 633–656.

Hughes, K. D., Higham, J. P., Allen, W. L., Elliot, A. J., and Hayden, B. Y. (2015). Extraneous red drives female macaques' gaze toward photographs of male conspecifics. *Evolution and Human Behavior* 36: 25–31.

Hurst, G. D., and Werren, J. H. (2001). The role of selfish genetic elements in eukaryotic evolution. *Nature Reviews Genetics* 2: 597–606.

Iannaccone, L. R. (1994). Why strict churches are strong. *American Journal of Sociology* 99: 1180–1211.

Igic, B., Cassey, P., Grim, T., Greenwood, D. R., Moskát, C., Rutila, J., and Hauber, M. E. (2012). A shared chemical basis of avian host–parasite egg colour mimicry. *Proceedings of the Royal Society B* 279: 1068–1076.

Irons, W. (2001). Religion as a hard-to-fake sign of commitment. In Nesse, R., ed., *The Evolution of Commitment*, 292–309. New York: Russell Sage Foundation.

Jacoby, S. (2004). *Employing Bureaucracy: Managers, Unions, and the Transformation of Work in the 20th Century.* Mahwah, NJ: Lawrence Erlbaum.

Jankowiak, W., Nell, M. D., and Buckmaster, A. (2002). Managing infidelity: A cross-cultural perspective. *Ethnology* 41: 85–101.

Janus, S., and Janus, C. L. (1993). *The Janus Report on Sexual Behavior.* Hoboken, NJ: John Wiley & Sons.

Jersakova, J., Johnson, S. D., and Kindlmann, P. (2006). Mechanisms and evolution of deceptive pollination in orchids. *Biological Reviews* 81: 219–235.

Jones, B. C., Hahn, A. C., and DeBruine, L. M. (2019.) Ovulation, sex hormones, and women's mating psychology. *Trends in Cognitive Sciences* 23: 51–62.

Jørgensen, T. B. (2012). Weber and Kafka: The rational and the enigmatic bureaucracy. *Public Administration* 90: 194–210.

Juslin, P. N., and Västfjäll, D. (2008). Emotional responses to music: The need to consider underlying mechanisms. *Behavioral and Brain Sciences* 31: 559–621.

Kamiya, S., Han Kim, Y., and Suh, J. (2016). The face of risk: CEO testosterone and risk taking behavior. Working Paper. Singapore: Nanyang Technological University.

Kant, I. (1797). On a Supposed Right to Lie from Altruistic Motives. In Beck, L. W., ed. and trans., *Critique of Practical Reason and Other Writings in Moral Philosophy.* New York: Bobbs-Merrill, 1956.

Kaptchuk, T. J., and Miller, F. G. (2015). Placebo effects in medicine. *New England Journal of Medicine* 373: 8–9.

Kar, D., and Freitas, S. (2011). Illicit Financial Flows from Developing Countries over the Decade Ending 2009. Global Financial Integrity (www.gfip.org).

Kaufmann, A. E. (2008). *Women in Management and Life Cycle: Aspects that Limit or Promote Getting to the Top.* New York: Palgrave Macmillan.

Kawai, K., Lang, R., and Li, H. (2018). Political kludges. *American Economic Journal: Microeconomics* 10: 131–158.

Kelley, L. A., and Endler, J. A. (2012). Illusions promote mating success in great bowerbirds. *Science* 335: 335–338.

Kelley, L. A., Coe, R. L., Madden, J. R., and Healy, S. D. (2008). Vocal mimicry in songbirds. *Animal Behaviour* 76: 521–528.

Kempenaers, B., and Schlicht, E. (2010). Extra-pair behaviour. In Kappeler, P. M., ed., *Animal Behaviour: Evolution and Mechanisms*, 359–412. Berlin: Springer.

Khare, A., and Shaulsky, G. (2010). Cheating by exploitation of developmental prestalk patterning in *Dictyostelium discoideum*. *PLoS Genetics* 2: e1000854.

Kiazad, K., Restubog, S. D., Zagenczyk, T. J., Kiewitz, C., and Tang, R. L. (2010). In pursuit of power: The role of authoritarian leadership in the relationship between supervisors' Machiavellianism and subordinates' perceptions of abusive supervisory behavior. *Journal of Research in Personality* 44: 512–519.

Kiers, E. T., Rousseau, R. A., West, S. A., and Denison, R. F. (2003). Host sanctions and the legume-rhizobium mutualism. *Nature* 425: 78–81.

Kikuchi, D. W., and Pfennig, D. W. (2010). Predator cognition permits imperfect coral snake mimicry. *American Naturalist* 176: 830–834.

Kojima, T., Oishi, K., Matsubara, Y., Uchiyama, Y., Fukushima, Y., Aoki, N., Sato, S. et al. (2019). Cows painted with zebra-like striping can avoid biting fly attack. *PLoS ONE* 14: e0223447.

Kokko, H., Brooks, R., McNamara, J. M., and Houston, A. I. (2002). The sexual selection continuum. *Proceedings of the Royal Society B* 269: 1331–1340.

Kruger, J., and Dunning, D. (1999). Unskilled and unaware of it: How difficulties in recognizing one's own incompetence lead to inflated self-assessments. *Journal of Personality and Social Psychology* 77: 1121–1134.

Krupenye, C., Kano, F., Hirata, S., Call, J., and Tomasello, M. (2016). Great apes anticipate that other individuals will act according to false beliefs. *Science* 354: 110–114.

Kumashiro, M., and Sedikides, C. (2005). Taking on board liability-focused feedback: Close positive relationships as a self-bolstering resource. *Psychological Science* 16: 732–739.

Kurup, R., Johnson, A. J., Sankar, S., Hussain, A. A., Sathish Kumar, C., and Sabulal, B. (2013). Fluorescent prey traps in carnivorous plants. *Plant Biology* 15: 611–615.

Kwan, V.S.Y., Barrios, V., Ganis, G., Gorman, J., Lange, C., Kumar, M., Shepard, A., and Keenan, J. P. (2007). Assessing the neural correlates of self-enhancement bias: A transcranial magnetic stimulation study. *Experimental Brain Research* 182: 379–385.

Larmuseau, M.H.D., Matthijs, K., and Wenseleers, T. (2016). Cuckolded fathers rare in human populations. *Trends in Ecology and Evolution* 31: 327–329.

Lefevre, C. E., Lewis, G. J., Perrett, D. I., and Penke, L. (2013). Telling facial metrics: Facial width is associated with testosterone levels in men. *Evolution and Human Behavior* 34: 273–279.

Levine, T. R. (2019). *Duped: Truth-default Theory and the Social Science of Lying and Deception*. Tuscaloosa: University of Alabama Press.

Lewis, M. (1993). The development of deception. In Lewis, M., and Saarni, C., eds., *Lying and Deception in Everyday Life*, 90–105. New York: Guilford Press.

Light, P. C. (2017). *People on People on People: The Continued Thickening of Government*. New York: The Volcker Alliance.

Linde, K., Streng, A., Jürgens, S., Hoppe, A., Brinkhaus, B., Witt, C., Wagenpfeil, S. et al. (2005). Acupuncture for patients with migraine: A randomized controlled trial. *Journal of the American Medical Association* 293: 2118–2125.

Linde, K., Witt, C. M., Streng, A., Weidenhammer, W., Wagenpfeil, S., Brinkhaus, B., Willich, S. N., and Melchart, D. (2007). The impact of patient expectations on outcomes in four randomized controlled trials of acupuncture in patients with chronic pain. *Pain* 128: 264–271.

Lindenfors, P., Nunn, C. L., and Barton, R. A. (2007). Primate brain architecture and selection in relation to sex. *BMC Biology* 5: 20.

Liu, D., Wei, R., Zhang, G., Yuan, H., Wang, Z.-P., Sun, L., Zhang, J.-X., and Zhang, H.-M. (2008). Male panda (*Ailuropoda melanoleuca*) urine contains kinship information. *Chinese Science Bulletin* 53: 2793–2800.

Lockwood, B. B., Nathanson, C. G., and Weyl, E. G. (2017). Taxation and the allocation of talent. *Journal of Political Economy* 125: 1635–1682.

Loftus, E. F., and Pickrell, J. E. (1995). The formation of false memories. *Psychiatric Annals* 25: 720–725.

Lopes, R. J., Johnson, J. D., Toomey, M. B., Ferreira, M. S., Araujo, P. M., Melo-Ferreira, J., Andersson, L., Hill, G. E., Corbo, J. C., and Carneiro, M. (2016). Genetic basis for red coloration in birds. *Current Biology* 26: 1427–1434.

Lusardi, A., and Mitchell, O. S. (2009). How ordinary consumers make complex economic decisions: Financial literacy and retirement reactions. NBER Working Paper 15350.

Lyle III, H. F., Smith, E. A., and Sullivan, R. J. (2009). Blood donations as costly signals of donor quality. *Journal of Evolutionary Psychology* 7: 263–286.

Lynn, M. (2010). *Bust: Greece, the Euro and the Sovereign Debt Crisis.* Hoboken, NJ: Bloomberg Press.

Lyon, B. E., and Eadie, J. M. (2008). Conspecific brood parasitism in birds: A life-history perspective. *Annual Review of Ecology, Evolution, and Systematics* 39: 343–363.

Macknik, S., Martinez-Conde, S., and Blakeslee, S. (2011). *Sleights of Mind: What the Neuroscience of Magic Reveals about Our Everyday Deceptions.* New York: Picador.

Mank, J. E., and Avise, J. C. (2006). Comparative phylogenetic analysis of male alternative reproductive tactics in ray-finned fishes. *Evolution* 60: 1311–1316.

Mason, R. T., Fales, H. M., Jones, T. H., Pannell, L. K., Chinn, J. W., and Crews, D. (1989). Sex pheromones in snakes. *Science* 245: 290–293.

Mather, M., and Carstensen, L. L. (2005). Aging and motivated cognition: The positivity effect in attention and memory. *Trends in Cognitive Science* 9: 496–502.

Maynard-Smith, J., and Harper, D. (2003). *Animal Signals.* Oxford: Oxford University Press.

McGeeney, B. E. (2015). Acupuncture is all placebo and here is why. *Headache Currents* 55: 465–469.

McLain, D. K., McBrayer, L. D., Pratt, A. E., and Moore, S. (2010). Performance capacity of fiddler crab males with regenerated versus original claws and success by claw type in territorial contests. *Ethology, Ecology, and Evolution* 22: 37–49.

McQuire, B., Olsen, B., Bemis, K. E., and Orantes, D. (2018). Urine marking in male domestic dogs: Honest or dishonest? *Journal of Zoology* 306: 163–170.

Meier, K. J., O'Toole, L., and Bohte, J. (2006). Inside the bureaucracy: Principals, agents, and bureaucratic strategy. In Meier, K., and O'Toole, L., eds., *Bureaucracy in a Democratic State*, 93–120. Baltimore: Johns Hopkins University Press.

Melville, P. (2014). Lying with Godwin and Kant: Truth and duty in *St. Leon*. *Eighteenth Century* 55: 19–37.

Merton, R. K. (1957). *Social Theory and Social Structure*. Glencoe, IL: Free Press.

Meyer, A. (2006). Repeating patterns of mimicry. *PLoS Biology* 4: e341.

Michener, C. D. (2000). *The Bees of the World*. Baltimore: Johns Hopkins University Press.

Møller, A. P. (1990). Deceptive use of alarm calls by male swallows, *Hirundo rustica*: A new paternity guard. *Behavioral Ecology* 1: 1–6.

Müller, F. (1879). *Ituna* and *Thyridia*; a remarkable case of mimicry in butterflies. (R. Meldola translation). *Proclamations of the Entomological Society of London* 1879: 20–29.

Müller-Schwarze, D. (2006). *Chemical Ecology of Vertebrates*. Cambridge: Cambridge University Press.

Mundy, N. I., Stapley, J., Bennison, C., Tucker, R., Twyman, H., Kim, K.-W., Burke, T., Birkhead, T. R., Andersson, S., and Slate, J. (2016). Red carotenoid coloration in the zebra finch is controlled by a cytochrome P450 gene cluster. *Current Biology* 26: 1435–1440.

Murphey, P. E., Laczniak, G. R., and Wood, G. (2007). An ethical basis for relationship marketing: A virtue ethics perspective. *European Journal of Marketing* 41: 37–57.

Myre, M. A. (2012). Clues to γ-secretase, huntingtin and Hirano body normal function using the model organism *Dictyostelium discoideum*. *Journal of Biomedical Sciences* 19: 41.

Neale, M. A., and Bazerman, M. H. (1985). The effects of framing and negotiator overconfidence on bargaining behaviors and outcomes. *Academy of Management Journal* 28: 34- 49.

Nelson, X. J. (2012). A predator's perspective of the accuracy of ant mimicry in spiders. *Psyche* 2012: 1–5.

Nelson, X. J., and Jackson, R. R. (2009). Collective Batesian mimicry of ant groups by aggregating spiders. *Animal Behaviour* 78: 123–129.

Niskanen, W. N. (1994). *Bureaucracy and Public Economics*. Fairfax, VA: The Locke Institute.

Nokelainen, O., Scott-Samuel, N. E., Nie, Y., Wei, F., and Caro, T. (2021). The giant panda is cryptic. *Scientific Reports* 11: 21287.

Nonacs, P. (2006). Nepotism and brood reliability in the suppression of worker reproduction in the eusocial Hymenoptera. *Biology Letters* 2: 577–579.

Norman, L. J., and Thaler, L. (2019). Retinotopic-like maps of spatial sound in primary "visual" cortex of blind human echolocators. *Proceedings of the Royal Society B* 286: 20191910.

Norris, G., Brookes, A., and Dowell, D. (2019). The psychology of Internet fraud victimisation: A systematic review. *Journal of Police and Criminal Psychology* 34: 231–245.

Norwitz, J. (2009). *Pirates, Terrorists, and Warlords: The History, Influence, and Future of Armed Groups around the World*. New York: Skyhorse.

Nunn, C. L., and Lewis, R. J. (2001). Cooperation and collective action in animal behavior. In Noë, R., van Hooff, J.A.R.A.M., Hammerstein, P., eds., *Economics in Nature*, 42–46. Cambridge: Cambridge University Press.

Nyberg, D. (1993). *The Varnished Truth: Truth Telling and Deceiving in Ordinary Life*. Chicago: Chicago University Press.

Nyhan, B., and Reifler, J. (2010). When corrections fail: The persistence of political misconceptions. *Political Behavior* 32: 303–330.

Odean, T. (1998). Volume, volatility, price, and profit when all traders are above average. *Journal of Finance* 53: 1887–1934.

Olson, D.V.A., and Perl, P. (2005). Free and cheap riding in strict, conservative churches. *Journal for the Scientific Study of Religion* 44: 123–142.

Onyishi, I. E., Prokop, P., Okafor, C. O., and Pham, M. N. (2016). Female genital cutting restricts sociosexuality among the Igbo people of southeast Nigeria. *Evolutionary Psychology* 14: 1–7.

Palazzo, A. F., and Gregory, T. R. (2014). The case for junk DNA. *PloS Genetics* 10: e1004351.

Parkinson, C. N. (1957). *Parkinson's Law*. Boston: Houghton Mifflin.

Paulhus, D. L., Harms, P. D., Bruce, M. N., and Lysy, D. C. (2003). The over-claiming technique: Measuring self-enhancement independent of ability. *Journal of Personality and Social Psychology* 84: 890–904.

Pazhoohi, F. (2016). On the practice of cultural clothing practices that conceal the eyes: An evolutionary perspective. *Evolution, Mind and Behaviour* 14: 55–64.

Pazhoohi, F., and Hosseinchari, M. (2014). Effects of religious veiling on Muslim men's attractiveness ratings of Muslim women. *Archives of Sexual Behavior* 43: 1083–1086.

Peter, L. J., and Hull, R. (1969). *The Peter Principle: Why Things Always Go Wrong*. New York: William Morrow.

Petersen, J. L., and Hyde, J. S. (2010). A meta-analytic review of research on gender differences in sexuality, 1993–2007. *Psychological Bulletin* 136: 21–38.

Plassmann, H., O'Doherty, J., Shiv, B., and Rangel, A. (2008). Marketing actions can modulate neural representations of experienced pleasantness. *Proceedings of the National Academy of Sciences* 105: 1050–1054.

Platek, S. M., Burch, R. L., Panyavin, I. S., Wasserman, B. H., and Gallup Jr., G. G. (2002). Reactions to children's face resemblance affects males more than females. *Evolution and Human Behavior* 23: 159–166.

Plath, M., Richter, S., Tiedemann, R., and Schlupp, I. (2008). Male fish deceive competitors about mating preferences. *Current Biology* 18: 1138–1141.

Pollard, K. A., and Blumstein, D. T. (2012). Evolving communicative complexity: Insights from rodents and beyond. *Philosophical Transactions of the Royal Society B* 367: 1869–1878.

Porter, S. S., and Simms, E. L. (2014). Selection for cheating across disparate environments in the legume-rhizobium mutualism. *Ecology Letters* 9: 1121–1129.

Powell, L. E., Isler, K., and Barton, R. A. (2017). Re-evaluating the link between brain size and behavioural ecology in primates. *Proceedings of the Royal Society B* 284: 20171765.

Price, D. D., Finniss, D. G., and Benedetti, F. (2008). A comprehensive review of the placebo effect: Recent advances and current thought. *Annual Review of Psychology* 59: 565–590.

Proctor, H. C. (1991). Courtship in the water mite *Neumartia papillator*: Males capitalize on female adoptions for predation. *Animal Behaviour* 42: 589–598.

Prum, R. O. (2018). *The Evolution of Beauty*. New York: Anchor.

Ratnieks, F.L.W., and Wenseleers, T. (2005). Policing insect societies. *Science* 307: 54–56.

Reddy, V. (2007). Getting back to the rough ground: Deception and "social living." *Philosophical Transactions of the Royal Society B* 362: 621–637.

Rice, W. R. (2013). Nothing in genetics makes sense except in light of genomic conflict. *Annual Review of Ecology, Evolution and Systematics* 44: 217–237.

Riebel, K., Odom, K. J., Langmore, N. E., and Hall, M. L. (2019). New insights from female bird song: Towards an integrated approach to studying male and female communication roles. *Biology Letters* 15: 20190059.

Riehl, C., and Frederickson, M. E. (2016). Cheating and punishment in cooperative animal societies. *Philosophical Transactions of the Royal Society B* 371: 20150090.

Rojas, B., Burdfield-Steel, E., de Pasqual, D., Gordon, S., Hernández, L., Mappes, J., Nokelainen, O., Rönkä, K., and Lindstedt, C. (2018). Multimodal aposematic signals and their emerging role in mate attraction. *Frontiers in Ecology and Evolution* 6: 93.

Rosenthal, G. G., and Evens, C. S. (1998). Female preference for swords in *Xiphophorus helleri* reflects a bias for large apparent size. *Proceedings of the National Academy of Sciences* 95: 4431–4436.

Rozenblit, L., and Keil, F. C. (2002). The misunderstood limits of folk science: An illusion of explanatory depth. *Cognitive Science* 26: 521–562.

Ryan, M. J., and A. S. Rand (1999). Phylogenetic influence on mating call preferences in female Túngara frogs, *Physalaemus pustulosus*. *Animal Behaviour* 57: 945–956.

Ryan, M. J., and Cummings, M. E. (2013). Perceptual biases and mate choice. *Annual Review of Ecology, Evolution, and Systematics* 44: 437–459.

Santorelli, L. A., Thompson, C.R.L., Villegas, E., Svetz, J., Dinh, C., Parikh, A., Sucgang, R. et al. (2008). Facultative cheater mutants reveal the genetic complexity of cooperation in social amoebae. *Nature* 451: 1107–1110.

Saul-Gershenz, L. S., and Millar, J. G. (2006). Phoretic nest parasites use sexual deception to obtain transport to their host's nest. *Proceedings of the National Academy of Sciences* 103: 14039–14044.

Scelza, B. A. (2011). Female choice and extra-pair paternity in a traditional human population. *Biology Letters* 7: 889–891.

Schaefer, H. M., and Ruxton, G. D. (2009). Deception in plants: Mimicry or perceptual exploitation? *Trends in Ecology and Evolution* 24: 676–685.

Schmidt, L., Skvortsova, V., Kullen, C., Weber, B., and Plassmann, H. (2017). How context alters value: The brain's valuation and affective regulation system link price cues to experienced taste pleasantness. *Scientific Reports* 7: 8098.

Schmitt, D. P., and Buss, D. M. (2001). Human mate poaching: Tactics and temptations for infiltrating existing relationships. *Journal of Personality and Social Psychology* 80: 894–917.

Schreiber, N., Bellah, L. D., Martinez, Y., McLaurin, K. A., Strok, R., Garven, S., and Wood, J. M. (2006). Suggestive interviewing in the McMartin Preschool and Kelly Michaels daycare abuse cases: A case study. *Social Influence* 1: 16–47.

Scott, D. J., Stohler, C. S., Egnatuk, C. M., Wang, H., Koeppe, R. A., and Zubieta, J. K. (2007). Individual differences in reward responding explain placebo-induced expectations and effects. *Neuron* 55: 325–336.

Scott-Phillips, T. C., Blythe, R. A., Gardner, A., and West, S. A. (2012). How do communication systems emerge? *Proceedings of the Royal Society B* 279: 1943–1949.

Sinervo, B., and Lively, C. M. (1996). The rock-paper-scissors game and evolution of alternative male strategies. *Nature* 380: 240–243.

Singer, N., Jacoby, N., Lin, T., Raz, G., Shpigelman, L., Gilam, G., Granot, R. Y., and Hendler, T. (2016). Common modulation of limbic network activation underlies musical emotions as they unfold. *Neuroimage* 141: 517–529.

Slocombe, K. E., and Zuberbühler, K. (2007). Chimpanzees modify recruitment screams as a function of audience composition. *Proceedings of the National Academy of Sciences* 104: 17228–17233.

Smith, D. L. (2007). *Why We Lie: The Evolutionary Roots of Deception and the Unconscious Mind.* New York: St. Martin's Griffin.

Sodian, B., and Frith, U. (1992). Deception and sabotage in autistic, retarded and normal children. *Journal of Child Psychology and Psychiatry* 33: 591–605.

Soler, M., Pérez-Contreras, T., and de Neve, L. (2014). Great spotted cuckoos frequently lay their eggs while their magpie host is incubating. *Ethology* 120: 965–972.

Sommer, V. (1994). Infanticide among the langurs of Jodhpur: Testing the sexual selection hypothesis with a long-term record. In Parmigiani, S., and vom Saal, F., eds., *Infanticide and Parental Care*, 155–198. Reading, UK: Harwood.

Sosis, R., and Alcorta, C. (2003). Signaling, solidarity, and the sacred: The evolution of religious behavior. *Evolutionary Anthropology* 12: 264–274.

Spottiswoode, C. N., and Koorevaar, J. (2012). A stab in the dark: Chick killing by brood parasitic honeyguides. *Biology Letters* 8: 241–244.

Steele, M. A., Halkin, S. L., Smallwood, P. D., McKenna, T. J., Mitsopoulos, K., and Beam, M. (2008). Cache protection strategies of a scatter-hoarding rodent: Do tree squirrels engage in behavioural deception? *Animal Behaviour* 75: 705–714.

Stegen, J. C., Gienger, C. M., Sun, L. (2004). The control of color change in the Pacific tree frog, *Hyla regilla. Canadian Journal of Zoology* 82: 889–896.

Steger, R., and Caldwell, R. L. (1983). Intraspecific deception by bluffing: A defense strategy of newly molted stomatopods (Arthropoda: Crustacea). *Science* 221: 558–560.

Stevens, M. (2016). *Cheats and Deceits: How Animals and Plants Exploit and Mislead.* Oxford: Oxford University Press.

Strassmann, B. I. (2003). Social monogamy in a human society: Marriage and reproductive success among the Dogon. In Reichard, U. H., and Boesch, C., eds., *Monogamy: Mating Strategies and Partnerships in Birds, Humans and Other Mammals*, 177–189. Cambridge: Cambridge University Press.

Strassmann, J. E., Zhu, Y., and Queller, D. C. (2000). Altruism and social cheating in the social amoeba *Dictyostelium discoideum. Nature* 408: 965–967.

Street, S. E., Navarrete, A. F., Reader, S. M., and Laland, K. N. (2017). Coevolution of cultural intelligence, extended life history, sociality, and brain size in primates. *Proceedings of the National Academy of Sciences* 114: 7908–7914.

Sullivan-Beckers, L., and Crocroft, R. B. (2010). The importance of female choice, male-male competition, and signal transmission as causes of selection on male mating signals. *Evolution* 64: 3158–3171.

Suls, J., Lemos, K., and Stewart, H. L. (2002). Self-esteem, construal, and comparisons with the self, friends, and peers. *Journal of Personality and Social Psychology* 82: 252–261.

Sun, C., Shepard, D. B., Chong, R. A., Arriaza, J. L., Hall, K., Castoe, T. A., Feschotte, C., Pollock, D. D., and Mueller, R. L. (2012). LTR retrotransposons contribute to genomic gigantism in plethodontid salamanders. *Genome Biology and Evolution* 4: 168–183.

Sun, L. (2003). Monogamy correlates, socioecological factors, and mating systems in beavers. In Reichard, U. H., and Boesch, C., eds., *Monogamy: Mating Strategies and Partnerships in Birds, Humans and Other Mammals*, 138–146. Cambridge: Cambridge University Press.

Sun, L., and Müller-Schwarze, D. (1998). Anal gland secretion codes for relatedness in the beaver, *Castor canadensis*. *Ethology* 104: 917–927.

Svenson, O. (1981). Are we all less risky and more skillful than our fellow drivers? *Acta Psychologica* 47: 143–148.

Syrůčková, A., Saveljev, A. P., Frosch, C., Durka, W., Savelyev, A. A., and Munclinge, P. (2015). Genetic relationships within colonies suggest genetic monogamy in the Eurasian beaver (*Castor fiber*). *Mammal Research* 60: 139–147.

Taglor, M. J. (2007). Deception (lying). In Baumeister, R. F., and Vohs, K. D., eds., *Encyclopedia of Social Psychology*, 220–221. Los Angeles: Sage.

Talwar, V., and Crossman, A. (2011). From little white lies to filthy liars: The evolution of honest and deception in young children. In Benson, J. B., ed., *Advances in Child Development and Behavior*, 40: 139–179. London: Academic Press.

Talwar, V., and Lee, K. (2008). Social and cognitive correlates of children's lying behavior. *Child Development* 79: 866–881.

Tamura, N. (1995). Postcopulatory mate guarding by vocalization in the Formosan squirrel. *Behavioral Ecology and Sociobiology* 36: 377–386.

Tanaka, K. D., and Ueda, K. (2005). Horsfield's hawk-cuckoo nestlings simulate multiple gapes for begging. *Science* 308: 653.

Tavris, C., and Aronson, E. (2015). *Mistakes Were Made (But Not by Me): Why We Justify Foolish Beliefs, Bad Decisions, and Hurtful Acts*. New York: Mariner Books.

Taylor, R. C., and Ryan, M. J. (2013). Interactions of multisensory components perceptually rescue Túngara frog mating signals. *Science* 341: 273–274.

Tenbmnsel, A. E. (1998). Misrepresentation and expectations of misrepresentation in an ethical dilemma: The role of incentives and temptation. *Academy of Management Journal* 41: 330–339.

Thaler, R. H., and Sunstein, C. R. (2008). *Nudge: Improving Decisions about Health, Wealth, and Happiness*. New Haven, CT: Yale University Press.

Tibbetts, E. A., and Izzo, M. (2010). Social punishment of dishonest signalers caused by mismatch between signal and behavior. *Current Biology* 20: 1637–1640.

Toma, C. L., and Hancock, J. T. (2010). Looks and lies: The role of physical attractiveness in online dating self-presentation and deception. *Communication Research* 37: 335–351.

Toye, J. (2006). Modern bureaucracy. WIDER Research Paper, No. 2006/52, United Nations University World Institute for Development Economics Research (UNU-WIDER), Helsinki.

Treas, J., and Giesen, D. (2000). Sexual infidelity among married and cohabiting Americans. *Journal of Marriage and the Family* 62: 48–60.

Trivers, R. (2011). *The Folly of Fools: The Logic of Deceit and Self-Deception in Human Life*. New York: Basic Books.

Turner, P. E. (2005). Cheating viruses and game theory: The theory of games can explain how viruses evolve when they compete against one another in a test of evolutionary fitness. *American Scientist* 93: 428–435.

Vallin, A., Jakobsson, S., Lind, J., and Wiklund, C. (2005). Prey survival by predator intimidation: An experimental study of peacock butterfly defence against blue tits. *Proceedings of the National Academy of Sciences* 272: 1203–1207.

Vallin, A., Jakobsson, S., and Wiklund, C. (2007). "An eye for an eye?": On the generality of the intimidating quality of eyespots in a butterfly and a hawkmoth. *Behavioral Ecology and Sociobiology* 61: 1419–1424.

Van der Linden, S., Leiserowitz, A., Rosenthal, S., and Maibach, E. (2017). Inoculating against misinformation. *Science* 358: 1141–1142.

Veblen, T. (1899). *The Theory of the Leisure Class: An Economic Study of Institutions.* New York: Penguin.

Vilmer, J.-B. J., Escorcia, A., Guillaume, M., and Herrera, J. (2018). Information manipulation: A challenge for our democracies. Report by the Policy Planning Staff (CAPS) of the Ministry for Europe and Foreign Affairs and the Institute for Strategic Research (IRSEM) of the Ministry for the Armed Forces, Paris.

Vnuk, A., Owen, H., and Plummer, J. (2006). Assessing proficiency in adult basic life support: Student and expert assessment and the impact of video recording. *Medical Teacher* 28: 429–434.

Von Hippel, W., and Trivers, R. (2011). The evolution and psychology of self-deception. *Behavioral and Brain Sciences* 34: 1–56.

Vosoughi, S., Roy, D., and Aral, S. (2018). The spread of true and false news online. *Science* 359: 1146–1151.

Wager, T. D., Rilling, J. K., Smith, E. E., Sokolik, A., Casey, K. L., Davidson, R. J., Kosslyn, S.M., Rose, R. M., and Cohen, J. D. (2004). Placebo-induced changes in FMRI in the anticipation and experience of pain. *Science* 303: 1162–1167.

Wallace, A. R. (1867). Mimicry and other protective resemblances among animals. *Westminster Review* 1–43.

Walton, J. P. (2019). *Twelve Lies that Hold America Captive: And the Truth that Sets Us Free.* Downers Grove, IL: IVP Book.

Walum, H., and Westberg, L. (2011). The behavioral genetics of human pair bonding. In Ebstein, R., Shamay-Tsoory, S., and Chew, S. H., eds., *DNA to Social Cognition*, 37–46. Hoboken, NJ: John Wiley & Sons.

Watts, D. P. (1989). Infanticide in mountain gorillas: New cases and a reconsideration of the evidence. *Ethology* 81: 1–18.

Weber, M. (1968/1921). *Economy and Society.* (Roth, G., and Wittich, C., eds.) New York: Bedminster.

Weinrib, J. (2008). The juridical significance of Kant's "Supposed Right to Lie." *Kantian Review* 13: 142–170.

Westen, D., Blagov, P. S., Harenski, K., Kilts, C., and Hamann, S. (2006). Neural bases of motivated reasoning: An fMRI study of emotional constraints on partisan political judgment in the 2004 U.S. presidential election. *Journal of Cognitive Neuroscience* 18: 1947–1958.

Westneat, D. F. (1987). Extra-pair copulations in a predominantly monogamous bird: Observations of behaviour. *Animal Behaviour* 35: 877–884.

Whisman, M. A., Gordon, K. C., and Chatav, Y. (2007). Predicting sexual infidelity in a population-based sample of married individuals. *Journal of Family Psychology* 21: 320–324.

Whiting, M. J., Webb, J. K., and Keogh, J. S. (2009). Flat lizard female mimics use sexual deception in visual but not chemical signals. *Proceedings of the Royal Society B* 276, 1585–1591.

Wickler, W. (1968). *Mimicry in Plants and Animals*. London: World University Library.

Wiederman, M. W. (1997). Extramarital sex: Prevalence and correlates in a national survey. *Journal of Sex Research* 34: 167–174.

Wilkinson, G. S. (1990). Food sharing in vampire bats. *Scientific American* 262: 64–70.

Wilson, D. S., Near, D., and Miller, R. R. (1996). Machiavellianism: A synthesis of the evolutionary and psychological literatures. *Psychological Bulletin* 119: 285–299.

Wood, C. (2016). Ritual well-being: Toward a social signaling model of religion and mental health. *Religion, Brain & Behavior* 7: 258–262.

Xu, F., Boa, X., Fu, G., Talwar, V., and Lee, K. (2010). Lying and truth-telling in children: From concept to action. *Child Development* 81: 581–596.

Yolles, M. (2016). Governance through political bureaucracy: An agency approach. *Kybernetes* 48: 7–34.

Zahavi, A. (1975). Mate selection: A selection for a handicap. *Journal of Theoretical Biology* 53: 205–214.

Zahavi, A., and Zahavi, A. (1997). *The Handicap Principle: A Missing Piece of Darwin's Puzzle*. New York: Oxford University Press.

Zhang, J.-X., Rao, X.-P., Sun, L., Zhao, C.-H., and Qin, X.-W. (2007). Putative chemical signals about sex, individuality, and genetic background in the preputial gland and urine of the house mouse (*Mus musculus*). *Chemical Senses* 32: 293–303.

Zhang, J.-X., Sun, L., Bruce, K. E., and Novotny, N. V. (2008). Chronic exposure of cat odor enhances aggression, urinary attractiveness and sex pheromones of mice. *Journal of Ethology* 26: 279–286.

Zhang, J.-X., Sun, L., and Novotny, M. (2007). Mice respond differently to urine and its major volatile constituents from male and female ferrets. *Journal of Chemical Ecology* 33: 603–612.

Zheng, Y. C., Yuan, T. T., and Liu, T. (2014). Is acupuncture a placebo therapy? *Complementary Therapies in Medicine* 22: 724–730.

Zuckerman, E. W., and Jost, J. T. (2001). What makes you think you're so popular? Self-evaluation maintenance and the subjective side of the "Friendship Paradox." *Social Psychology Quarterly* 64: 207–223.

Zuk, M., Rotenberry, J. T., and Tinghitella, R. M. (2006). Silent night: Adaptive disappearance of a sexual signal in a parasitized population of field crickets. *Biology Letters* 2: 521–524.

Zupancic, A. (2000). *Ethics of the Real: Kant, Lacan*. London: Verso.

INDEX